Philosophy Beyond Spacetime

Philosophy Beyond Spacetime

Implications from Quantum Gravity

Edited by

CHRISTIAN WÜTHRICH,
BAPTISTE LE BIHAN,
and
NICK HUGGETT

OXFORD
UNIVERSITY PRESS

OXFORD
UNIVERSITY PRESS

Great Clarendon Street, Oxford, OX2 6DP,
United Kingdom

Oxford University Press is a department of the University of Oxford.
It furthers the University's objective of excellence in research, scholarship,
and education by publishing worldwide. Oxford is a registered trade mark of
Oxford University Press in the UK and in certain other countries

First Edition published in 2021

Impression: 1

Published in the United States of America by Oxford University Press
198 Madison Avenue, New York, NY 10016, United States of America

British Library Cataloguing in Publication Data
Data available

Library of Congress Control Number: 2021937972

ISBN 978–0–19–884414–3

DOI: 10.1093/oso/9780198844143.001.0001

Printed and bound by
CPI Group (UK) Ltd, Croydon, CR0 4YY

Contents

Acknowledgements

We are grateful to everyone who interacted with us in the context of our project *Space and Time After Quantum Gravity*. All the speakers and participants at the events we hosted contributed immensely to the project's success. We owe special thanks to the contributors of this collection. In particular, we would like to acknowledge the winners of our essay contests: in this volume the outstanding chapters by Adam Koberinski and Tushar Menon. The interdisciplinary nature of the collection required significant efforts of intellectual openness in addressing audiences and in working from perspectives that differed for many from their usual work. We also thank our assistants for their help and the contests' anonymous judges for their valuable feedback and considered recommendations. The final preparation of this volume also owes a lot to our editorial assistant at the University of Geneva, Gaia Valenti, whose help was invaluable. We thank the staff at OUP, in particular Peter Momtchiloff, Céline Louasli, Kalpana Sagayanathan, and Sally Evans-Darby for their help in the production of the collection. Finally, we acknowledge the financial support from the John Templeton Foundation for making the project and this collection possible.

List of Figures

List of Contributors

Jeremy Butterfield, Trinity College, University of Cambridge

David J. Chalmers, Department of Philosophy, New York University

Richard Healey, Department of Philosophy, University of Arizona

Nick Huggett, Department of Philosophy, University of Illinois at Chicago

Jenann Ismael, Department of Philosophy, Columbia University

Adam Koberinski, Department of Philosophy, University of Waterloo

Baptiste Le Bihan, Department of Philosophy, University of Geneva

Kerry McKenzie, Department of Philosophy, University of California at San Diego

Tushar Menon, Department of Philosophy, University of Cambridge

Alyssa Ney, Department of Philosophy, University of California at Davis

Daniele Oriti, Arnold Sommerfeld Center for Theoretical Physics, Ludwig Maximilian University of Munich

Alastair Wilson, Department of Philosophy, University of Birmingham

Christian Wüthrich, Department of Philosophy, University of Geneva

David Yates, Centro de Filosofia, Faculdade de Letras, Universidade de Lisboa

1
Introduction

Christian Wüthrich, Baptiste Le Bihan, and Nick Huggett

Abstract

Quantum gravity offers a fertile ground for philosophical work, particularly through its suggestion that spacetime may not be fundamental but merely a derivative structure. As such, theories of quantum gravity stand in a long tradition of physical theories with deep implications for the nature of space and time, and indeed the fundamental structure of our material world. This Introduction summarizes the contributions to this collection by structuring them around three themes. The first group of chapters analyses various aspects of the search of lost spacetime in quantum gravity. The second group studies metaphysical and epistemological aspects of the emergence in play in quantum gravity. The third group widens the investigations to several key methodological challenges arising in the context of quantum gravity.

Contemporary physics has much to teach us about the nature of space and time, as has become obvious with the advent of relativity theory at the latest. What is less obvious is that the relativistic revolution may only have been the first step in a longer process of deconstructing our pre-theoretical categories of space and time. In fact, attempts to unify the lessons of general relativity (GR) with the other great revolution of the twentieth century in physics, quantum physics, into a theory of quantum gravity (QG) suggest a rather strange idea: that *space and time as we know them do not fundamentally exist, but instead emerge from a non-spatiotemporal structure*. Thus, the physics under construction, which we hope will one day provide a more unified and fundamental view of reality, could lead to a novel understanding of the nature of space and time, radically opposed to everything previously believed. This volume is the result of our conviction that this stunning consequence, difficult as it may be to conceptualize, could genuinely change the way in which many discussions in philosophy are conducted—say, on the existence of space, the flow of time, or the boundaries of space and time, to name but a few examples. Hence the title of this collection: *Philosophy Beyond Spacetime*.

Christian Wüthrich, Baptiste Le Bihan, and Nick Huggett, *Introduction* In: *Philosophy Beyond Spacetime: Implications from Quantum Gravity*. Edited by: Christian Wüthrich, Baptiste Le Bihan, and Nick Huggett, Oxford University Press.
 DOI: 10.1093/oso/9780198844143.003.0001

The current situation in contemporary physics results from a long historical journey. As we have already mentioned, two revolutions took place in physics during the twentieth century: special and then general relativity, and quantum physics. The first gave a new understanding of gravity as a feature of the geometry of space and time, with dramatic philosophical consequences: the relativity of simultaneity, the dynamical relation between spacetime and matter, the possibility of temporal anomalies, and the existence of singularities in spacetime, for example. The second deals with the matter that inhabits spacetime, and poses equally deep philosophical challenges: the non-local connection between entangled systems, the appearance of a classical world, the very existence of particles. While being of enormous empirical power and accuracy, both have raised deep and unresolved philosophical and foundational questions—as the ever-expanding literature on the subjects attests.

There is as yet no generally accepted unified theory of quantum matter and gravity, capable of describing all phenomena that need to be understood. Quantum matter cannot easily be described as interacting with classical relativistic spacetime. But it seems as if matter and geometry must be closer in nature than they are in the two separate theories. By itself, this observation gives us reason to think that a successful theory of QG will raise similarly important challenges for our conceptions of space, time, and matter—perhaps abolishing them altogether as fundamental entities. Although such a theory does not yet exist, we may nevertheless see the silhouettes of coming changes in the fragmentary theories that do exist, and come to understand new possibilities for epistemology and metaphysics. But just as important, there is good reason to think that some of the problems in finding a theory of QG are themselves conceptual, in need of philosophical analysis by philosophers and physicists—just as relativity required Einstein's reconceptualizing of the nature of space, time, and motion, and as quantum mechanics was driven by the competing pictures of Schrödinger and Heisenberg.

This volume is one of the fruits of a three-year research project, *Space and Time after Quantum Gravity* (2015–18), funded by the John Templeton Foundation. The project involved both physicists working on the physics of QG and philosophers working on this topic or related notions in contemporary philosophy. We have selected a few outstanding works, the fruit of intense discussions over several years, and we have divided them into two volumes. One—*Beyond Spacetime: The Foundations of Quantum Gravity* (Cambridge University Press, 2020)—deals more with philosophical questions arising in the technical development of different approaches to QG. The other volume—this one—is more directly concerned with the implications of QG for questions traditionally seen as more philosophical in nature. Of course, this distinction is gradual at best and often blurred—not the least in foundational research in QG. Several of the chapters of either collection could equally well have been included in either volume. To a first degree of approximation, the companion volume, which requires more technical skills in physics, addresses a wider range of physicists, while we hope to reach many philosophers beyond the narrow confines of technically demanding philosophy of physics with the present volume.

The key aim of this volume is to expand knowledge and understanding of the philosophy of QG by the philosophical community. It emphasizes how debates in metaphysics—regarding time, emergence, composition, or grounding, for example—shed light on the conceptual questions of QG; and conversely, how quantum theories of space and time call into question philosophical views grounded in classical spacetime. Furthermore, the philosophy of QG raises methodological questions, for instance concerning the relation between physics and metaphysics. The essays in this volume have been chosen to demonstrate to a wide range of philosophers the significance of the subject, as well as making novel contributions to it.

The essays are organized around three main subjects: (i) the possible emergence of spacetime in various approaches to QG, (ii) philosophical (especially metaphysical and epistemological) discussions of the nature of this relation of emergence, and (iii) a final section devoted to methodological aspects of the philosophy of QG. The remainder of this Introduction sketches these topics and the contributions.

1.1 Searching for Spacetime

The first series of chapters explores various approaches to QG and examines how spacetime might emerge in these specific approaches. The first chapter categorizes the way in which spacetime can be said to emerge from a more fundamental but less spatiotemporal structure into qualitatively different levels and exemplifies them in the context of loop quantum gravity (LQG). The first three chapters all aim to clarify, and perhaps to some extent downplay, claims of emergence of familiar spacetime structures in this context. The last chapter, in contrast, proposes that the classical notion of spacetime would have to be generalized if nature turns out to be supersymmetric.

The first chapter by **Daniele Oriti** proposes a levels view of the emergence of spacetime in QG. According to Oriti, the emergence of spacetime comes in degrees, where 'degree' or 'level' is not to be understood ontologically as metaphysical layers or a sequential succession. Rather, levels are intended to indicate a broadening of the perspective on the problem, such that at each step, novel conceptual, methodological, epistemological, or ontological issues show up as the complexity and the richness of the physics increase. He identifies four distinct levels of emergence of spacetime and of the gravitational field, and offers the helpful analogy of the emergence of hydrodynamic properties of (super)fluids from its atomic constitution.

The first level, Level 0, covers classical GR and direct quantizations of its geometric degrees of freedom. Already at the classical level, there is a sense in which space and time dissolve, as the general covariance of classical GR implies that, generically, there exists a multitude of notions of space and time. Moreover, in the Hamiltonian formulation of GR required for canonical quantization, there is a 'problem of time' in that time and dynamics appear to vanish altogether. Even so, the fundamental degrees of freedom are still spatiotemporal, or geometric. These difficulties are exacerbated at the quantum level, where superpositions of exact geometric configurations are permitted. In the hydrodynamical analogy, Level 0 would correspond to the construction of macroscopic hydrodynamic observables as functions of different ones, with their quantization adding difficulties. Following Carlo Rovelli, Oriti believes that these challenges can be resolved by deploying a relational strategy.

Level 1 complicates this picture by adding new, distinct kinds of degrees of freedom, which are neither spatiotemporal nor geometrical in any direct sense. Typically, the fundamental degrees of freedom are combinatorial or algebraic, and show no continuum structure. Many theories of QG, such as LQG, string theory, causal sets, and causal dynamical triangulations, are naturally interpreted to at least ascend to this level. While the explication of emergence at Level 0 involves the classical limit, to see spacetime emerge at Level 1, the continuum limit—to be carefully distinguished from the classical limit—must be taken (usually via coarse graining or renormalization). Unlike the other levels, Level 1 includes an ontological aspect in that the postulated new kinds of degrees of freedom form a novel ontological category. This level is directly analogous to the move from macroscopic hydrodynamics to the grainy world of the atoms which constitute the fluid.

Levels 2 and 3 do not proceed to a novel, qualitatively distinct kind of degrees of freedom. Instead, they involve different conceptual perspectives on the same basic ontology. Level 2 starts from the realization that the continuum limit is often not unique, but leads to distinct macroscopic phases separated by possible phase transitions. Not all such macroscopic phases will be geometric or spatiotemporal. What needs to be shown to resolve this level is that there exists a spatiotemporal phase in some approximation and some limit. In the case of such non-trivial macroscopic phase diagrams, spacetime, argues Oriti, can be said to be emergent in a more radical sense compared to Level 1. In the hydrodynamic analogy, one and the same system can condense into different macroscopic phases such as solid, fluid, and gaseous, and will only obey hydrodynamic laws in one of them.

Finally, the existence of distinct macroscopic phases leads to the possibility that the system undergoes a phase transition. Thus, a system may transition from a non-spatiotemporal to a spatiotemporal or geometric phase, sustaining a 'geometrogenesis'. This is what may occur in quantum models of the big bang where one can distinguish an 'earlier', non-spatiotemporal phase from a 'later', approximately spatiotemporal one. While this transition ought to be regarded as a physical process, the difficulties involved in conceptualizing these transitions as physical yet non-temporal processes justify regarding them as Level 3 transitions. In comparison, appreciating the physical nature of phase transitions in the hydrodynamic analogy seems more straightforward.

Oriti's categorization of qualitatively different kinds of spacetime emergence gives useful guidance to a field which has so far only started to appreciate the systematically different kinds of issues involved at the different level of emergence.

Jeremy Butterfield's chapter addresses the nature and significance of 'dualities' in physics. These symmetries attracted a great deal of interest amongst physicists when they were discovered in string theory in the 1990s, sparking the 'second string revolution'. More recently they have been the subject of intense scrutiny by philosophers, for their implications for the interpretation of string theory (for a general overview aimed at philosophers, see Le Bihan and Read 2018). For instance, Huggett (2017) argues that duals should be understood as fully equivalent descriptions, so that any apparent differences are non-factual. If so, the T-duality between theories with different radii for the universe means that fundamentally space has no definite radius, and, developing a line of thought proposed by Brandenberger and Vafa (1989), the observed definite radius must be emergent.

Butterfield argues that such reasoning should be resisted. Drawing on work by (and with) Sebastian De Haro (e.g. 2020), he first gives a formal account of 'duality': broadly speaking, two theories are dual when they are different formal representations of a common 'bare theory'. For instance, one can represent the same system of moving bodies in frames of reference with different origins, orientations, and states of rest. There is an isomorphism between the two representations that allows one to see how they are describing a common set of quantities in the same way. However, this formal mapping says nothing by itself about whether one of the representations is right about absolute rest, or whether there is in fact no standard of absolute rest.

Such questions are a matter of the interpretation of physical theory, in this case the two duals. Do they automatically say the same things, or disagree substantively? Butterfield (and De Haro) therefore invoke a formal theory of interpretation, in which reference is handled by actual worldly extension, and sense by extension across possible worlds. Of course, in this framework, whether the duals say the same things depends on their intended sense: Newton's 'space' has an absolute standard of rest, which Newton hypothesized to be that of 'the centre of the world'. Of course, in that sense of 'space', one could meaningfully disagree with him, even though Galilean symmetry means that no experiment can settle the issue. Similarly for T-duals: one could in principle understand their difference as either merely representational or as real. For Butterfield, until we have a theory that explicitly unifies them, revealing the underlying bare theory, we must understand them as disagreeing about the actual world. Those advocating for emergence will disagree, arguing that one can in suitable circumstances infer that duals agree factually, even absent an explicit underlying theory (e.g. Huggett and Wüthrich 2020).

In her chapter, **Alyssa Ney** asks whether there is evidence for the emergence of spacetime—specifically metrical—structure from quantum entanglement entropy: this is an idea that has received recent attention (for instance, Cao et al. 2017). A starting point for this way of thinking (though related ideas go back further) is the Ryu-Takayanagi conjecture that 'holographically' relates the entanglement on the boundary of anti-de Sitter spacetime to an area in the bulk, using the AdS-CFT duality. (Specifically, the entropy arising from the entanglement between the conformal fields in two regions on the boundary is directly related to the area of the minimal surface in the bulk that separates them.) It is quite striking that two things as disparate as quantum entanglement—which at root measures the ability to factorize vectors in Hilbert space—and metrical quantities can be related in this way. It is an insight that led Cao and co-authors to describe how to reconstruct a spatial metric from an abstract quantum system.

Such an idea strongly suggests emergence. The quantum structure, from which the spatial structure is supposedly recovered, is seemingly non-spatial, as we noted. However, as Ney explains (drawing on the earlier work of others), that conclusion is hasty, for the mere correspondence does not entail emergence. Amongst other possibilities, which, if either, of the two sides is more fundamental? The correspondence alone does not tell us that it is the Hilbert space side. If, Ney argues, one looks at the derivation of the Ryu-Takayanagi conjecture, the reasoning seems, if anything, to show that it reflects the way spatial structure constrains the quantum entanglement. (And, she claims, similarly in earlier derivations of the Hawking-Bekenstein entropy that also relate entropy and area: e.g. Bombelli et al. 1986.) In that way of looking at things, space seems to be at the same level as entanglement. Thus, she concludes, these derivations of space from Hilbert space do not support the claim of emergence. Of course, that is not to say that they are incorrect, or that there may not be other motivations for seeking to derive space in this way: for instance, one might hope that by starting from such a solidly quantum foundation, one might eventually recover emergent GR, thereby providing a route to a full theory of QG. But Ney's claim is that such motivations are future-oriented speculations, not at present well supported by metric-entanglement correspondence.

According to what **Tushar Menon** calls 'Earman's principle', dynamical and spacetime symmetries ought to coincide in a theory. Thus, if the dynamics of matter is Lorentz covariant, as is the case for the standard model, then the appropriate spacetime venue for the theory is one which is Lorentz symmetric, i.e. Minkowski spacetime. If it turned out to be the case that matter enjoys an additional, and hitherto undetected, symmetry, the spacetime should follow suit and be generalized accordingly.

In his contribution, Menon explores the possibility that matter is supersymmetric, which is the case if the theory's Lagrange density remains invariant under transformations mapping bosons to fermions and vice versa. As it turns out, such transformations 'mix' with spacetime translations in the sense that repeated applications result in a net translation in spacetime. Menon concludes from this that supersymmetry is inherently spatiotemporal, giving us all the more reason to extend the spacetime arena for supersymmetric physics. Such a superspace consists in a 'supermanifold'—constructed from commuting and anticommuting 'supernumbers'—endowed with a 'super-Minkowski metric'. The reward for formulating supersymmetric physics in superspace is that the resulting equations of motion are manifestly supersymmetric, allowing the theory to wear its symmetry on its sleeve.

As it turns out, the light postulate of special relativity is violated in such a supertheory: the speed of light is no longer invariant in all superspace coordinate systems. Menon concludes from this that we should be hesitant to read off the operational information concerning the behaviour of measuring devices such as rods and clocks made from supersymmetric fields from the structure of superspace. Maybe so, but this might alternatively be counted as a strike against the extension of Earman's principle to the supersymmetric context.

1.2 The Metaphysics of Spacetime Emergence

That space and time or spacetime might emerge from a non-spatiotemporal structure seems frankly puzzling—and difficult to articulate conceptually. Hence the second series of selected chapters, devoted to the philosophical analysis of the claim that spacetime is emergent. The series comprises four chapters written by David Yates, David J. Chalmers, Alastair Wilson, and Jenann Ismael, respectively. They discuss the prospects of analysing the dependence of spacetime on a non-spatiotemporal structure via a *relation of functional realization* (an important milestone in a debate which is gaining momentum—cf. the forthcoming special issue of *Synthese* on the topic, edited by Karen Crowther, Niels Linnemann, and Christian Wüthrich), a *relation of causation*, or even the possibility to *fully eliminate spacetime from the furniture of the world.*

In his contribution, **David Yates** examines the problem of empirical incoherence in QG: if spacetime is not part of the fundamental ontology of physics, how is it possible for fundamental physical theories to be justified by observations of spatiotemporally located things like rods, pointers, and clocks? As a solution to the problem, he distinguishes between views that accept and reject the reality of a spatiotemporal structure over and above the more fundamental non-spatiotemporal structure. Drawing lessons from the philosophy of mind, he argues that we should adopt a form of realism about the dependent spatiotemporal structure in order to solve the problem of empirical incoherence.

This dependent spacetime, according to Yates, can either be grounded in or caused by the more fundamental non-spatiotemporal structure. Thus, Yates sees spacetime as a set of functional roles implemented by a more fundamental structure, a view called 'spacetime functionalism', which was developed in the context of QG by Lam and Wüthrich (2018). The spatiotemporally located things that act as evidence for physical theories are thereby regarded as being real entities caused by, or grounded in, more fundamental non-spatiotemporal entities. Yates thereby makes a case for a specific sort of spacetime functionalism that takes the existence of spatiotemporal roles, ontologically speaking, very seriously—against another sort of spacetime functionalism, more linguistic in nature, that regards spatiotemporal roles as mere linguistic roles some predicates occupy in the architecture of the theory. The view is at odds with Linnemann (forthcoming) who argues that spacetime functionalism as a general view does not help with the problem of empirical incoherence.

David J. Chalmers focuses first on the emergence of space rather than of spacetime with a focus on philosophy of mind and the possibility of identifying what he calls 'Edenic space'—namely, space as we immediately find it in the manifest image of the world—with real physical space. His main claim is that we should be functionalists about the manifest image, and that we should not expect the Edenic space of the manifest image to faithfully mirror the structure of the physical space. Rather, we should construe space as the structure triggering (our experience of) the manifest spatial image. In other words, Chalmers argues against what he dubs 'spatial primitivism', namely the view that space is just as it appears in the manifest image. He favours instead *spatial functionalism*—the view that space is what causes the Edenic space of the manifest image.

The move from spatial primitivism to spatial functionalism has major fallouts, not only for the status of the physical space in which we live but also for other sorts of more abstract spaces involved in virtual reality experiences. Indeed, he argues that the rejection of spatial primitivism in favour of spatial functionalism has a straightforward consequence: virtual spaces should be regarded as real spaces, functionally implemented by an ontology to which we have no direct epistemic access via our experience of the virtual reality. Then he moves on to analysing the interpretation of spacetime in the context of GR and of the relation of emergence existing between the relativistic spacetime and the more fundamental non-spatiotemporal structure of QG. Likewise and in line with other works, he discusses the strategy of finding spatiotemporal functional roles in the general theory of relativity before turning to quantum theories of gravity to identify the realizers of these roles. There again, he defends the view that the functionalist strategy justifies realism about the spacetime of GR. Drawing from discussions about the existence of an explanatory gap between a non-spatiotemporal theory of QG and a theory of GR, Chalmers then focuses on the possible existence of another explanatory gap, this time between GR and the manifest image; he argues that spacetime functionalism may close this gap. In brief, according to him, we should free our ontology from Edenic space assumptions; but we should nonetheless be realists about a functional physical space, identified with the structure that triggers our experience of Edenic space.

Alastair Wilson also examines spacetime functionalism as a way to analyse the dependence of spacetime on a non-spatiotemporal structure. He begins with the following question: should we construe the dependence of spacetime on the non-spatiotemporal structure as causal or non-causal? As Wilson points out, distinguishing between causal and non-causal relations—also called 'grounding relations'—requires a demarcation criterion. This criterion is supposed to allow us to decide, in the face of a relation of dependency, whether this relation is causal or not. An intuitive demarcation criterion is related to time. Causation would refer to cross-temporal dependency relations—the two relata of the relation being located at different times—and grounding to synchronic dependency relations—the two relata being located at the very same time. As Wilson notes, if this temporal demarcation principle were to be right, then it would follow that the emergence of spacetime cannot be causal (and not grounding either), since time doesn't exist at the fundamental level—thereby preventing us from identifying the dependence of spacetime on the non-spatiotemporal structure to a causal relation.

But Wilson takes another road as, drawing on previous work (Wilson 2018), he rejects the temporal criterion and proposes an alternative mediation criterion: causal relations are dependence relations mediated by a law of nature, while grounding relations are mediated by constitutive principles—what it is to be a particular thing or kind of thing. This suggests that we can use our understanding of what it is to be a law of nature to grasp the distinction between causation and grounding. Of course, as Wilson acknowledges, this amounts to replacing questions about the nature of dependency relations with questions over the nature of those mediating principles. For instance, if we construe causal relations as relations mediated by a law of nature, then we need to understand what a law of nature is—without appealing to causation, at risk of vicious circularity—which might or might not be problematic, depending on which of the two notions one takes to be more fundamental. Nonetheless, it is an important result that the difference between grounding and causation can be illuminated this way as we can then analyse, in some approaches, emergent spacetime as being caused by a non-spatiotemporal structure.

As Wilson notes, this idea might surprise the reader as, at first sight, it seems at odds with the common assumption found in the literature that spacetime does not emerge via a causal process, causation being too tightly related to the existence of space and time (a possible exception could be found in cosmological models based on QG; see e.g. Huggett and Wüthrich 2018). Usually, this relation of dependency is considered to be both non-temporal and non-causal, either because it is a primitive relation of grounding (i.e. not to be specified further via another relation) or because it is a relation of mereological composition. But Wilson's new criterion suggests that spacetime could literally be *caused by* a non-spatiotemporal structure, in particular in LQG. Wilson then shows that there is a way to avoid this surprising consequence by using the functionalist machinery. Spacetime functionalism offers a new way to interpret the situation in LQG, spacetime being functionally realized—a constitutive principle—and hence grounded in, rather than caused by, the non-spatiotemporal structure, according to the mediation criterion. Wilson closes with a discussion of the many-instant landscape view defended by Gomes (2017). According to this view, the world is a 'timeless' state space of spatial field configurations. Wilson argues again that the existence of spacetime in this approach should be regarded as mediated by a functional principle and so, because of his demarcation principle, as grounded in the non-spatiotemporal structure.

In her chapter, **Jenann Ismael** argues for an expansive view of what the 'emergence of spacetime' might amount to, which opens an unnoticed and perhaps surprising route to recovering spacetime. It is typically assumed that conscious experience will be recovered from fairly high-level physics, so that reduction from any lower physics can (with a sigh of relief) ignore the mind—to be taken care of later. Ismael cites Maudlin (2007) as an example of this position. Since the physics which is supposed to explain the mind is spatiotemporal, the corollary is that physical space itself will have to be recovered from any non-spatiotemporal theory—as programmes in QG typically assume.

However, of course all that need be recovered, strictly speaking, is the conscious experience of spatiotemporality. If one assumes that that is given immediately in experience, then the result is the same—physical spacetime must emerge. But, Ismael argues, spacetime is not at all immediately experienced. Empirical studies support the view—which has antecedents in Berkeley, Poincaré, and Mach (on whom she focuses)—and provide evidence that the concept of spacetime is a construct from experience. More specifically, different sensory modalities present different perspectives on the physical world, each in its own 'sensory manifold'. But the mind unifies these through sensory and motor interaction with the world, finding that a representation in which they are viewed as different takes on a single spatiotemporal world simplifies their relations greatly. (The significant plasticity in this unification shows that it is not a fixed given.)

The upshot of course is that strictly speaking, a theory without spacetime is not required to deliver spacetime itself, but only to explain the sensory modalities, to recover the sensory manifolds. Then the brain itself will take care of spacetime, which would be recognized as a mere appearance. Ismael notes that such an account would fit a functionalist model: the functions of the modalities could be found to be played by the non-spatiotemporal. She does not necessarily advocate such a strategy, but argues that it should be recognized to understand the epistemology and goals of spacetime emergence. At first glance the strategy appears radical, but Ismael points out that it may not be so far from what has already been suggested. For instance, Rovelli (1991) suggests a strategy of recovering rods and clocks instead of spacetime: it is not such a great leap to consider instead recovering 'information gathering and utilising machines', cashed out functionally in non-spatiotemporal terms (see also Baron 2020).

1.3 Methodological Issues

The last series of chapters reflects the diversity of works in the growing community of philosophy of QG. If a central issue in QG is the conceptual articulation of the potential emergence of spacetime, many other issues in QG arise along the way. With this last section, we want to invite philosophers to delve deeper into those other questions related to the construction of a theory of QG—and thus to draw their attention to a number of points that could have important repercussions for many philosophical questions.

The first chapter by Richard Healey deals with the perhaps most famous problem in the foundations of quantum mechanics, *the measurement problem*. However, it does so in the context of one specific approach to quantum gravity, LQG, raising the difficult question of whether or not we should solve the measurement problem in order to make progress with QG. The second chapter by Kerry McKenzie focuses on the connection between physics and philosophy, and on the possibility of obtaining metaphysical knowledge from the current state of physics where we lack, one must note, a final theory applying to all domains of observation. The final chapter by Adam Koberinski examines a deep puzzle about vacuum energy, sometimes called the 'vacuum catastrophe' or 'cosmological constant problem'. The problem arises from a clash between the value of the vacuum energy we calculate via quantum field theory—which must be extremely high—and the value we get from cosmological data when we observe the way the energy is distributed in the observable universe, suggesting an extremely low value.

These three issues, although quite different, may serve to show just how fertile research in the philosophical foundations of QG is for very different questions, traditional and new, technical and fundamental.

Richard Healey's contribution discusses critically, but sympathetically, Carlo Rovelli's relational interpretation (1996) as a way to solve the quantum measurement problem in the context of (covariant) LQG. Although Rovelli was of course one of the main architects of this theory, Healey shows that there are significant tensions between the two. Healey believes that these can be overcome in his own pragmatist approach to quantum physics.

Rovelli's relational interpretation of quantum mechanics departs from von Neumann's notion of a measurement resulting in a correlation between the initial state of the 'system' and the final state of the 'measurement device'. In Rovelli's interpretation, there is no place for an absolute and objective state of a system; instead, all states are relational in that a system is in a state only relative to another system, which may be an observer who has or gains knowledge about the first system. Consequently, Rovelli rejects the idea that there is such a thing as the complete description of the total state of the world.

Applying the relational interpretation to covariant LQG, Rovelli and Vidotto (2015) propose to understand its transitions between spin network states as physical processes enclosed between interactions between systems whose boundaries are ultimately conventional. It is these *processes* which are ultimately nothing but spacetime regions. The resulting 'relational loop quantum gravity', Healey argues, struggles to accommodate the concept of an observer capable of registering the outcomes of measurement interactions as it deals exclusively in spacetime regions (and their conventional boundaries). Since observers are not mere spacetime regions, more work is required to show how they (and thus von Neumann measurements) can be modelled in the context of relational LQG.

As one of the morals to be drawn, Healey concludes that the characterization of observing or measuring systems needs to be established as emergent in LQG. In this sense, he requires the emergence of spacetime for the quantum measurement to be resolved. As he notes, this is in disagreement with Wüthrich (2017), who argues instead that the emergence of spacetime in QG requires the resolution of the measurement problem, not the other way around.

The second chapter of this series, by **Kerry McKenzie**, begins with the observation that many philosophers dream of a final physical theory of everything that will answer some of the deepest and most intriguing metaphysical questions. For instance, is time really flowing or is it just a perceptual artefact of the way we experience a static four-dimensional world? Is there a sort of modal glue connecting events in a systematic way and explaining why the world seems to obey some laws of nature? We do not have such a final, absolutely fundamental theory, but this situation does not prevent metaphysicians from engaging in metaphysical activity by using constraints from empirically well-confirmed physics.

McKenzie argues that this kind of naturalized metaphysics is problematic. If we step back and look at theoretical physics as a whole, we can see that the empirically well-confirmed theories on which it is based should ideally be replaced by another more fundamental, and perhaps definitive, theory. Until we find this Holy Grail—assuming that it exists—we are stuck in a far from ideal epistemological situation. Indeed, it seems difficult to reliably draw metaphysical knowledge from the current state of physics. Why should we trust our currently most fundamental physical theories to give us empirical access to the fundamental structure of the world? After all, we know that those theories are not absolutely fundamental.

After characterizing her preferred sense of 'naturalistic metaphysics', McKenzie shows that claims in metaphysics had to change their truth value as science, and especially physics, progressed. With regard to the classic debate between scientific realists and anti-realists, she argues that friends of scientific realism may appeal to the notion of approximate truth and argue that certain structures are preserved by the change in theory and, therefore, can be seen to justify certain claims about the world. McKenzie goes on to argue that this notion of approximate truth is of little use in dealing with more metaphysical issues. Indeed, metaphysical questions (say, regarding the existence of past and future entities) are often ontological or substantive questions that are difficult to answer in terms of approximation. McKenzie concludes that the value of engaging in metaphysical speculation based on our currently most fundamental physical theories is unclear.

We believe that this situation offers a powerful motivation to shift our attention from the well-established theories of physics to QG. Indeed, McKenzie's challenge to naturalized metaphysics does not necessarily oblige us to wait for the development and empirical confirmation of a final theory. Le Bihan (2020), for example, argues in response to McKenzie that we can obtain substantial metaphysical knowledge from speculative physics by examining the field of QG as a whole and looking for features present in all or almost all approaches to QG.

The final chapter, by **Adam Koberinski**, engages with a topic at the interface of QG and cosmology. The notorious 'cosmological constant problem' concerns what some physicists have labelled the 'worst prediction in physics of all times': theoretical expectations for the value of the cosmological constant Λ based on considerations from particle physics overshoot observational limits anywhere from 50 to 120 orders of magnitude (depending on a choice of 'cutoff'). The problem arises because quantum field theory seems to suggest that empty space has an enormous energy density, which, according to GR, ought to manifest itself in the geometry of spacetime, coupling to the Einstein equation as a cosmological constant term, inflating the latter to gargantuan proportions inconsistent with observations.

After clarifying what exactly the problem is supposed to be, Koberinski argues that this argument ought to be resisted at each turn. First, he shows that there is little basis on which to accept an objective zero-point energy scale in quantum field theory and hence to take the vacuum energy as real. Second, even if one did accept this, the vacuum energy turns out to be badly divergent on standard renormalization procedures and hence does not deserve to be trusted. Third, it turns out that there exist more rigorous ways of coupling quantum field theory to GR than is assumed in the standard argument to the cosmological constant problem, and that under these approaches, vacuum energy, and so the cosmological constant problem, does not arise in the first place.

In the last part of his contribution, he shows that even assuming that all these steps to the cosmological constant problem can be justified, the presently dominant attempts to solve it all fail. These attempts include 'naturalness' approaches such as supersymmetry, apparent violations of the equivalence principle (for example, due to higher dimensions as we find them in string theory), or statistical avenues based on anthropic reasoning or quantum statistical considerations. All of this leads Koberinski to the sobering conclusion that the cosmological constant problem has not been established as an actual problem, and physicists taking it as a heuristic to develop new physics may well be barking up the wrong tree.[1]

References

Baron, S. (2020). The curious case of spacetime emergence. *Philosophical Studies 177*, 2207–2226.

Bombelli, L., R. K. Koul, J. Lee, and R. D. Sorkin (1986). Quantum source of entropy for black holes. *Physical Review D 34*(2), 373–383.

Brandenberger, R. and C. Vafa (1989). Superstrings in the early universe. *Nuclear Physics B 316*(2), 391–410.

Cao, C., S. M. Carroll, and S. Michalakis (2017). Space from Hilbert space: Recovering geometry from bulk entanglement. *Physical Review D 95*(2), 024031.

Crowther, K., N. S. Linnemann, and C. Wüthrich (forthcoming). Spacetime functionalism in general relativity and quantum gravity. *Synthese*. https://doi.org/10.1007/s11229-020-02722-z.

De Haro, S. (2020). Spacetime and physical equivalence. In N. Huggett, K. Matsubara, and C. Wüthrich (Eds.), *Beyond Spacetime: The Foundations of Quantum Gravity*, pp. 257–283. Cambridge: Cambridge University Press.

[1] Grant number 56314 from the John Templeton Foundation, performed under a collaborative agreement between the University of Illinois at Chicago and the University of Geneva. The contents of the work produced under this grant are solely the responsibility of the authors and do not necessarily represent the official views of the John Templeton Foundation.

Gomes, H. (2017). Quantum gravity in timeless configuration space. *Classical and Quantum Gravity 34*(23), 235004.

Huggett, N. (2017). Target space ≠ space. *Studies in History and Philosophy of Modern Physics 59*, 81–88.

Huggett, N. and C. Wüthrich (2018). The (a)temporal emergence of spacetime. *Philosophy of Science 85*(5), 1190–1203.

Huggett, N. and C. Wüthrich (2020). Out of nowhere: Duality. arXiv preprint arXiv:2005.12728. Forthcoming in *Out of Nowhere: The Emergence of Spacetime in Quantum Theories of Gravity*.

Huggett, N. and C. Wüthrich (forthcoming). *Out of Nowhere: The Emergence of Spacetime in Quantum Theories of Gravity*. Oxford: Oxford University Press.

Lam, V. and C. Wüthrich (2018). Spacetime is as spacetime does. *Studies in History and Philosophy of Modern Physics 64*, 39–51.

Le Bihan, B. (2020). String theory, loop quantum gravity and eternalism. *European Journal for Philosophy of Science 10*(2), 1–22.

Le Bihan, B. and J. Read (2018). Duality and ontology. *Philosophy Compass 13*(12), e12555.

Linnemann, N. (forthcoming). On the empirical coherence and the spatiotemporal gap problem in quantum gravity: And why functionalism does not (have to) help. *Synthese*. https://doi.org/10.1007/s11229-020-02659-3.

Maudlin, T. W. (2007). Completeness, supervenience and ontology. *Journal of Physics A: Mathematical and Theoretical 40*(12), 3151–3171.

Rovelli, C. (1991). What is observable in classical and quantum gravity? *Classical and Quantum Gravity 8*(2), 297–316.

Rovelli, C. (1996). Relational quantum mechanics. *International Journal of Theoretical Physics 35*(8), 1637–1678.

Rovelli, C. and F. Vidotto (2014). *Covariant Loop Quantum Gravity: An Elementary Introduction to Quantum Gravity and Spinfoam Theory*. Cambridge: Cambridge University Press.

Wilson, A. (2018). Metaphysical causation. *Noûs 52*(4), 723–751.

Wüthrich, C. (2017). Raiders of the lost spacetime. In D. Lehmkuhl, G. Schiemann, and E. Scholz (Eds.), *Towards a Theory of Spacetime Theories*, pp. 297–335. New York: Springer.

2
Levels of Spacetime Emergence in Quantum Gravity

Daniele Oriti

Abstract

We explore the issue of spacetime emergence in quantum gravity by articulating several levels at which this can be intended. These levels correspond to the reconstruction moves that are needed to recover the classical and continuum notion of space and time, which are progressively lost in a progressively deeper sense in the more fundamental quantum gravity description. They can also be understood as successive steps in a process of widening of the perspective, revealing new details and new questions at each step. Each level carries indeed new technical issues and opportunities, and raises new conceptual issues. This deepens the scope of the debate on the nature of spacetime, both philosophically and physically.

2.1 Introduction

The problem of quantum gravity is terribly multi-faceted and can be characterized in very different ways. It is the problem of obtaining a quantum theory of geometry and spacetime, a complete quantum description of gravitational phenomena. This is the common understanding: a possibly modified version of quantum mechanics should provide the mathematical framework of the theory and the object of the theory should be the gravitational interaction and the geometry of spacetime. The latter have been inextricably linked by General Relativity (GR) and nobody expects this link to be eliminated in a more fundamental quantum gravity theory. Beyond this common understanding one finds a variety of perspectives, which is moreover rapidly changing over time. This variety of perspectives, in turn, corresponds to a variety of approaches (Oriti 2009). One could identify two main schemes, each comprising several specific formalisms: one corresponding to the idea of quantum gravity resulting from quantizing a classical theory of geometry and gravity (e.g. GR) (Teitelboim 1982; Thiemann 2001; Kiefer 2006), the other in which spacetime,

Daniele Oriti, *Levels of Spacetime Emergence in Quantum Gravity* In: *Philosophy Beyond Spacetime: Implications from Quantum Gravity.* Edited by: Christian Wüthrich, Baptiste Le Bihan, and Nick Huggett, Oxford University Press.
 DOI: 10.1093/oso/9780198844143.003.0002

geometry, and gravity are in some sense 'emergent' from something else (Seiberg 2007; Oriti 2014; Padmanabhan 2015). In fact, not only is the distinction very coarse grained, but it is ambiguous since the issue of the 'emergence' of features of spacetime and geometry appears also in the first scheme. The emergent paradigm is the most recent and it is acquiring traction in recent years. Especially from the perspective of this second scheme, the problem of quantum gravity can be stated as: to identify the fundamental (quantum) degrees of freedom of spacetime, the 'atoms' of space (or spacetime); to define a consistent quantum dynamics for them; to show that a continuum and classical spacetime (with a geometric and matter fields) emerges from it, in some approximation; to show that GR is a good effective description of the dynamics of this emergent spacetime.

Quantum gravity in general, and the emergent paradigm in particular, face a large number of conceptual issues and raise an even larger number of philosophical questions (Isham 1991; Butterfield and Isham 2001; Kiefer 2013). This is inevitable, given the fundamental nature of the problem, shaking the very foundations of our thinking about the natural world, i.e. space and time. The (necessary and useful; see Oriti 2019) existence of a number of different approaches tackling the problem from different conceptual perspectives makes the situation more complex still. Plus, every solution is tentative, and every approach is incomplete, even when solid or promising. We are truly at the chaotic frontier of knowledge. The situation for philosophical reflections is excellent. It is also very different, however, from most philosophy of physics, since we are not dealing with the conceptual issues arising within established (mathematically and observationally) physical theories. The only way to deal with this peculiar situation is to exercise extra caution in adopting the points of view coming from specific approaches to quantum gravity as if they were more established than they are, and to refrain from resting too much on specific results as if they were a necessary part of any future theory. The same situation calls for more work to map this complex landscape (see also Mielczarek and Trześniewski 2018), especially at the conceptual level. This is what we hope to achieve with our contribution: a tentative map of the meanings in which space and time can be understood as 'emergent' in quantum gravity, and of the conceptual issues associated to this emergence, and thus a greater conceptual clarity about these issues.

The notion of emergence is itself subtle to define, even in ordinary physical theories (Batterman 2006, 2011; Bedau and Humphreys 2008). We will base our analysis on a very general characterization of it, provided by Butterfield and collaborators (Butterfield 2011a,b; Butterfield and Bouatta 2012). Emergence is understood to be the appearance, in a certain description of a physical system, of properties that are *novel* with respect to a different (more 'fundamental') description of the same system, and *robust*, thus stable enough to represent a characterization of the new description and to form part of new predictions stemming from it. Emergence, in this understanding, usually requires the use of some *limiting procedure* and of a number of (possibly drastic) *approximations*, to allow the novel properties to become visible in the new description.

This notion of emergence is compatible with the situation in quantum gravity, and it has been indeed already applied in this context (Butterfield 2011a; Huggett and Wüthrich 2013; Oriti 2014). Our analysis will be based on this and on a growing literature about the emergence of space and time in quantum gravity, concerning both physical and epistemological issues, among which: how to characterize this emergence and which physical consequences it may (or may not) have (Maudlin 2007; Bain 2013; Huggett and Wüthrich 2013; Crowther 2014; Huggett and Wüthrich 2018; Lam and Wüthrich 2018; Wüthrich 2019), what are the ontological implications of emergent spacetime scenarios (Lam and Wüthrich 2013; Wüthrich 2020), and more. Like the rest of the philosophy of quantum gravity, reflections on these issues could impact considerably, we believe, on philosophy of physics more generally, and on metaphysics and epistemology, since they challenge important aspects of these domains as well.

The scope and content of this contribution, however, are much more limited.

We will illustrate four levels of emergence for space, time, and geometry (thus, the gravitational field) in quantum gravity formalisms. We discuss four ways in which space, time, and geometry may be said to disappear in quantum gravity and, consequently, have to emerge to recover the description provided by GR, within a more fundamental quantum gravity formalism. These four levels have to be understood as successive steps in a process of widening of the perspective, revealing new details and new conceptual issues and new questions at each step. They also represent a deepening of our understanding of the issue of the emergence of space and time in quantum gravity. They should not be misunderstood as successive, sequential ontological, or inter-theoretical steps. They are not characterized each by different entities and they are not described each by a different theoretical framework. On the contrary, some of them can share the same fundamental degrees of freedom and all can be part of the same theoretical framework or quantum gravity formalism.

Let us clarify further the way we see our four levels of emergence,[1] in order to make the following presentation clearer.

First, the word 'levels' suggests an ordering among them. Such ordering should be understood in the following sense. At each step, as said, new issues emerge and new questions need to be tackled. In other words, what increases in going

[1] We use the word (and the concept of) 'emergence' in the sense clarified above, which we maintain applies to all levels we will identify in the following. This being said, different and more specific notions of emergence, as also proposed in the literature, could also be applied to the same four levels, but will most likely apply only to some of them or differently in each.

from one level to the next is the complexity and the richness of the required understanding of spacetime and, correspondingly, of the understanding of its emergence in quantum gravity. One could maybe speak instead of 'degrees' of emergence, indicating again such complexity and richness of understanding and of corresponding physical/mathematical description. This increased complexity or richness does not correspond directly to energy/distance scales (both are in fact spacetime-dependent notions, thus inapplicable in this context) or to metaphysical layers (of different fundamental vs. emergent entities, which only apply to the step from level 0 to level 1, as we will see). Thus, the 'levels of emergence' we speak about correspond instead to epistemological steps that we are required to take for a progressively improved understanding of the issue of the nature and emergence of spacetime, the improvement itself being *defined* exactly by the mentioned greater complexity and richness.

Second, we have also characterized the move from one level to the next as a deepening of our understanding of spacetime nature and emergence. Also, the notion of 'depth' of understanding that we adopt, and thus the idea of progress hinted at above, necessitates some clarification. Our working definition is the following. An issue is considered as understood in a deeper manner and a problem is considered as tackled at a deeper level if they are 'dressed' by a more complex network of related sub-issues and questions (the latter becoming visible, so to speak, at 'higher resolution', as a finer-grained conceptual visualization of the same issue is adopted). This does not mean that the deeper level is necessarily more complicated than the more superficial one, in terms of 'relevant entities' (metaphysical) or 'explanatory principles' (epistemological). On the contrary, very often the converse applies: simpler constituents may give rise to a plethora of complex emergent phenomena, just as simple general explanatory principles may be declined in a rich variety of concrete manners, depending on specific contexts. On the other hand, even when the latter situations arise, we maintain that the new 'level of understanding' obtained by discovering simpler constituent entities or simpler explanatory principles is more complex, and therefore deeper, because of the increased complexity of their consequences, i.e. because the *overall picture* has become richer.[2]

Finally, we may also characterize each successive level of spacetime emergence as more 'fundamental' than the previous, in line with the usual characterization of emergence in physical theories. It requires, though, an important caveat, following from what we said above. Only level 1 is more fundamental than level 0 in our classification, as we will explain, if one refers to the ontology corresponding to

[2] It may be interesting to develop this notion of depth of understanding, in the context of scientific theories, in more detail and relate it to the notion of 'explanatory depth' as discussed, for example, in Strevens (2008). We leave this to future work.

each level in terms of primary/derived relations. Level 2 and level 3 are on equal ontological footing as level 1, since the entities they refer to are the same. What changes is their understanding/characterization and the number and nature of questions referring to them that are tackled at each level. If instead by 'more fundamental' one refers to the improvement of our understanding of the world, i.e. a better understanding is dubbed a more fundamental understanding of it, then indeed, each step from one level to the next in our classification is a step towards a more fundamental level of (understanding of) reality. We also point out that the use of such labels implies, implicitly, a picture of progress in science as motion in the direction of increased complexity and refinement of understanding, in the sense specified above. This is another aspect that would deserve further work.

2.2 Level 0: Classical and Quantum (Modified) General Relativity

The zeroth level of spacetime emergence is the one corresponding to the traditional idea of quantum gravity as 'quantized GR' (or variations thereof). Quantum geometrodynamics and (Euclidean) quantum gravity path integrals à la Misner-Hawking, as well as canonical loop quantum gravity as initially understood (and still pursued by many), are examples of such traditional ideas. Quantum supergravity studied as a (non-)perturbative quantum field theory is another example, and the asymptotic safety scenario would be another, if the metric field is maintained as the fundamental entity.

In the classical theory, we have a covariant set of equations for the spacetime metric (identified with the gravitational field) and matter fields living on the same differentiable manifold, following from the gravitational action of choice. These equations encode the dynamics of spacetime.

The latter can be identified with the metric field itself or with the spatiotemporal quantities (temporal intervals, spatial distances, etc.) computed out of it. Since material objects are usually required to give physical meaning to such quantities, one can instead identify spacetime with specific combinations of matter and metric fields. One could call spacetime also the differentiable manifold itself (after all, this is what gives the first intuitive notion of 'spacetime point'), but this is of dubious physical significance, since the dependence of physical quantities on individual points in the differentiable manifold is removed by the request for diffeomorphism invariance (Rovelli and Gaul 2000), the gauge symmetry of GR.

Diffeomorphism invariance is indeed a key mathematical ingredient at the root of many of the conceptual difficulties about the nature of space and time in classical GR, and which have to do with the variety of possible identifications hinted at above (Norton 2003; Pooley 2010).

A more physical way of characterizing these difficulties is to say that they arise from the fact that every ingredient of the theory entering the definition of 'spacetime', matter, and metric fields is *dynamical* and that the dynamics itself, and its generic solutions, do not select any preferred time or space direction. On the contrary, the theory admits an infinity of equally valid local notions of time and space (that could be associated to specific coordinate frames, but without attributing to the latter any special physical significance).

In this sense, one can already speak of a disappearance of space and time in classical relativistic gravitational theory (Rovelli 2020). It is bypassed, in some sense, by the use of special solutions of the dynamics, which possess global spacetime symmetries and thus select special spacetime directions. In fact, much of gravitational physics rests on the use of such solutions.

At the conceptual level, this is already a big challenge to our customary conceptions of space and time, and raises many subtle issues, which form the subject of a vast literature in the philosophy of spacetime (Earman 1989; Albert 2000; Dainton 2001).

Notice that we are not distinguishing, here, between Lagrangian or Hamiltonian formulations of the theory, even though they are not strictly speaking equivalent (diffeomorphism symmetry is implemented in subtly different ways in the two settings, and a canonical Hamiltonian formulation requires global hyperbolicity, thus it is a priori less general than the covariant, Lagrangian one). We are also not distinguishing between space and time, even though the absence of a preferred notion of time is especially troublesome for our usual understanding of physical dynamics and of physics more generally. These special difficulties are the 'problem of time' in classical GR. These distinctions are not crucial to the main points we want to make in this contribution.

At the quantum level, assuming a standard formulation of quantum mechanics, the situation is much the same, just a little worse. The kinematics has states forming a Hilbert space which encode the geometry (intrinsic and extrinsic) of spatial submanifolds (possibly forming the boundaries of the differentiable manifold) and the values and momenta of matter fields on them, and the possible histories of the same states, thus the spacetime metric and the matter fields for the whole manifold. The dynamics is encoded in some operator equation, taking necessarily the form of a constraint equation for the same data, or in a sum-over-histories, path-integral formulation of the 'transition amplitudes' or '2-point functions' between the same quantum states (e.g. those interpreted as defining a physical scalar product for them), depending on the chosen classical action. Again, and for the same reasons as in the classical case, no preferred space or time direction is present in the theory, coordinate frames are unphysical, and generic physical configurations of the quantum spacetime will also not select any. The situation is worse than in the classical theory because even quantum states that solve the dynamics and possess special symmetries will usually not select exact metric or matter configurations, but mostly because a preferred time direction is essential to the standard formulation and interpretation of quantum theory itself, for any physical system we know of. Thus, in the quantum gravity case we are at a loss. However, this seems to us more an important problem in the foundations of quantum mechanics (can we build a consistent interpretation of quantum mechanics that does not rely, even implicitly, on a notion of time?) that any quantum gravity theory will force us to tackle, rather than new problems with the nature of space and time themselves, which remain essentially those of the classical theory.

At both classical and quantum levels, a solution to the problem of time and space can be found in the relational strategy (Rovelli 1991b; Gambini and Porto 2001). It takes on board the main lesson of GR, and it rephrases it in a way that immediately suggests a tentative solution: there is no time and no space, but only physical (imperfect) clocks and rods. The strategy amounts to identifying internal degrees of freedom of the complete system composed of metric and matter fields that can be used as *approximate* rods and clocks to parametrize the spatial relations and temporal evolution of the remaining degrees of freedom. To us, this is an adequate solution[3] to the issue of defining space and time in a (quantum) relativistic context, and a very physical one (but for a sample of the remaining issues, see Page and Wootters 1983; Rovelli 1991a; Gambini and Porto 2001). It forces us, however, to accept the fact that physical clocks and physical rods will never be perfect, i.e. matching the idealized (but unphysical) notion of time and space provided by coordinate systems. This is simply the other inevitable side of their being physical systems: quantum and interacting with the ones they parametrize.

There is thus a sense in which space and time disappear in classical GR and, in a more drastic sense, in the quantum GR. There is thus also a sense in which space and time have to *emerge* also in this context. In the classical case, this amounts to the dynamical selection of symmetric spacetimes or to the approximation leading to physical rods and clocks behaving as perfect ones. In the quantum case it is part of the standard problem of the classical approximation of a quantum theory, since the above symmetric spacetimes or geometries, and the close-to-ideal clocks and rods, have to emerge from ultimately quantum entities. General covariance, once more, leads to several additional complications to this notoriously already difficult problem, but, we maintain, does not change its nature. Most of the above

[3] To be clear, we do not want to imply at all that the relational strategy is uncontroversial or that it enjoys full consensus in the quantum gravity community. Different perspectives are taken. Beside the formalisms or perspectives in which full diffeomorphism invariance is renounced and a global notion of time is invoked in correspondence with preferred foliations of spacetime, one can try to extract a notion of time evolution, for example, from any (causal) ordering relation between events or boundary configurations in a diffeomorphism-invariant path integral.

challenges are well explored in the quantum gravity literature. But this is only level 0 of spacetime emergence.

2.3 Level 1: New Degrees of Freedom—Geometry and Spacetime as Emergent Entities

An altogether different sense in which space and time disappear in quantum gravity, and thus have to emerge in some approximation, is central in quantum gravity approaches that do not deal simply with quantized gravitational and matter fields. A new level is reached when quantum gravity formalisms are based on new types of quantum degrees of freedom which are not geometric in a straightforward way, but of a different nature, usually combinatorial and algebraic. In particular, this often implies a fundamental discreteness of the same quantum entities. The spin network states of loop quantum gravity (Ashtekar and Lewandowski 2004; Perez 2013; Bodendorfer 2016), with their dual functional dependence on group elements or group representations associated to graphs, and their histories labelled by the same algebraic data and associated to cellular complexes, fit this characterization.[4] The simplicial (piecewise-flat, thus singular) geometries of lattice quantum gravity approaches like quantum Regge calculus (Hamber 2009) and (causal) dynamical triangulations (Ambjørn, Görlich, Jurkiewicz, and Loll 2012) can also be understood in this perspective. The quanta of group field theories (GFTs) (Krajewski 2013; Oriti 2012), which can be described both as generalized spin networks and as simplicial building blocks of piecewise-flat geometries, and whose quantum dynamics merges the idea of spin foam models and that of lattice quantum gravity, are another example. Causal sets (Dowker 2013) are another purely discrete replacement for continuum fields. String theory offers a number of results all pointing to the replacement of the notion of continuum geometric fields as fundamental entities (Blau and Theisen 2009) and to a much more general type of geometry being reconstructed from the dynamics of strings (Hohm, Lüst, and Zwiebach 2013). Other examples could be cited. The main point should be clear: in quantum gravity, the fundamental degrees of freedom are not continuum fields and spacetime dissolves into pre-geometric, non-spatiotemporal entities, from which space, time, and geometry have to emerge in some approximation.

With the appearance of new fundamental (quantum) entities replacing continuum fields, call them 'atoms of space', an altogether new dimension opens up for quantum gravity research. Besides identifying the properties and dynamics of such fundamental entities, the crucial task becomes understanding by which

[4] This is true even though, historically, they have been 'discovered' within a rather conservative strategy of quantizing the gravitational field once it has been recast in the language of gauge theories.

physical mechanisms and under which approximations they become amenable to a description in terms of continuum spacetime and geometry (and matter fields). This is the problem of the continuum limit in discrete quantum gravity approaches. It must be carefully distinguished, conceptually and mathematically, from the classical limit mentioned in the previous section. It rests on coarse graining and renormalization schemes, the identification of appropriate collective observables, and in particular the identification of the usual continuum fields (metric, matter fields) as examples of collective quantities built out of the more fundamental atoms of space.

It is along this new dimension that we expect space and time to emerge in a stronger sense from entities that are not spatiotemporal. The new physics leading to the emergence of spacetime, more precisely, can be expected to be the one captured by increasing numbers of interacting, quantum fundamental entities, with space, time, and geometry arising from the collective behaviour of the same. Collective behaviour is indeed the prototypical producer of emergent physics, and the conceptual setup would here see spacetime in analogy with condensed matter and quantum many-body systems (Oriti 2018). With increasing numbers of fundamental entities, there come new emergent properties, new approximations, and effective dynamics, and with them new concepts.

The above expansion of the scope and content of quantum gravity research has been discussed in some detail in Oriti (2020), including also a brief survey of recent developments. These include spin foam lattice renormalization (Dittrich 2012; Bahr, Dittrich, Hellmann, and Kaminski 2013; Bahr and Steinhaus 2017; Delcamp and Dittrich 2017; Dittrich 2017), continuum limits in random tensor models (Gurau and Ryan 2012; Rivasseau 2016) and dynamical triangulations (Ambjørn, Görlich, Jurkiewicz, and Loll 2012), GFT renormalization (Geloun and Rivasseau 2013; Carrozza, Oriti, and Rivasseau 2014; Benedetti, Geloun, and Oriti 2015; Carrozza 2016; Geloun, Martini, and Oriti 2016; Carrozza, Lahoche, and Oriti 2017), and the extraction of effective cosmological dynamics as GFT hydrodynamics (Gielen, Oriti, and Sindoni 2013; de Cesare, Pithis, and Sakellariadou 2016; Gielen and Sindoni 2016; Oriti, Sindoni, and Wilson-Ewing 2016; Oriti 2017; Gielen and Oriti 2018).

Here, we want to stress the conceptual shift that such expansion brings, concerning the nature of space and time. When looked at from the point of view of the (candidate) pre-geometric atoms of space (level 1), it is clear that space, time, geometry, and matter have dissolved in a deeper, more radical sense than from the perspective of quantized GR (level 0). There, one had to get rid of the idea of any preferred notion of space and time, and a multiplicity of potential physical notions (be they defined by special configurations of spacetime geometry or by relationally defined frames). All such notions were made possible and had to be constructed by means of continuum fields. They are now missing. Thus, even the *possibility of space and time*, in the sense of level 0, has yet to emerge. It has to be obtained by moving along the new dimension of increasing numbers of fundamental building blocks and by exploring their collective properties.

This raises a number of questions concerning the nature of the atoms of space themselves. In particular, to what extent do they carry spatiotemporal properties at all? The need to reconstruct space and time from them, at least in some approximation and with respect to special aspects of their collective dynamics, implies that they carry at least 'seeds' of space and time with them. By this we mean that some of their properties should be *translatable* into spatiotemporal notions at least in those approximations, even though they are not fully spatiotemporal in general. In other words, if spacetime has to be reconstructed at all, the more fundamental theory should allow for a dictionary, mapping its basic entities and some of their properties into continuum fields including those defining spatiotemporal notions, in some sector of the same theory and in an approximate manner. The map will certainly not be one to one, or exact, but it should exist if the candidate fundamental theory is to have any physical relevance at all. The existence of such a dictionary, i.e. being part of the domain of this translation map, implies a 'proto-spatiotemporal' characterization of some properties of the fundamental atoms of space (and justifies their name). Nothing more than that should be assumed, however.

A more precise characterization requires probably to consider specific examples of candidate atoms of space and of quantum gravity formalisms. In particular, it is possible that some properties attributed to such atoms of space, among those that are crucial in reconstructing the standard notions of space and time, can be understood as offering a more primitive notion of space and time (e.g. based on adjacency, ordering, etc.), farther away from usual physics, but arguably more fundamental. A more primitive spatiotemporal reality would then replace, despite its radical departure from any traditional understanding (and use) of space and time, the one that we are accustomed to. This may end up being simply a matter of nomenclature. If the new properties are truly radically different (in mathematical and physical understanding) from the space and time of continuum relativistic physics, to call them 'spatiotemporal' may not offer more than a psychological relief.

An important issue is the ontological nature of the new fundamental entities underlying spacetime and, conversely, of space and time themselves, once we deprive them of their fundamental status and understand them as emergent. In fact, modern ontology (van Inwagen and Sullivan 2018) is based explicitly or implicitly on spatiotemporal notions, to the point that 'to be real' is often thought equivalent to 'to exist in space and time', i.e. to have a well-defined location and stable duration. This already raises ontological issues concerning the fields (in particular the metric field) that are used to define location and duration in relativistic physics. But the same ontological issues are brought to a whole new stage when referred to the putative atoms of space, underlying the same continuum fields and replacing them at the fundamental level. Conversely, unless one adopts the radical opposite of the usual position (to be real is to exist in space and time) and thus deprives space and time of *any reality at all*, due to their loss of fundamental status, one is forced to revise the very notion of reality in the presence of emergent behaviour. One has to accept a multi-level ontology of some sort, in which both fundamental and emergent properties and entities are real in an appropriate sense. In other words, an emergent spacetime scenario forces a radical revision of metaphysics in parallel with the revolution in physics that it represents, concerning what is meant by *real* (which has to be independent to some extent from spatiotemporal properties) and what this attribute is assigned to (which probably has to be done in a more liberal and less exclusive way). For recent work on these issues, see Butterfield (2011a,b); Butterfield and Bouatta (2012); Lam and Esfeld (2013); Wüthrich (2020).

Another set of issues raised by emergent spacetime scenarios is of a more epistemological nature. It concerns the physical salience of the candidate atoms of space and of the theories describing them, and their empirical coherence (Maudlin 2007; Huggett and Wüthrich 2013; Oriti 2014). The worry is that, because we live in spacetime and the notions of space (e.g. locality) and time (e.g. duration) are at the very root of our empirical access to reality, any theory formulated without them is either empirically empty or empirically incoherent. We maintain (Oriti 2014) that the necessary requirement of reproducing some (possibly modified) form of relativistic spacetime physics settles the worry of empirical emptiness of emergent spacetime scenarios, at least as a matter of principle. We also maintain that the empirical coherence of the same scenarios will have to be ensured by the details of such spacetime reconstruction, and of course tested in each specific formalism, but again that there is no obstruction in principle. The conceptual difficulties of course remain, and have to be consistently and seriously tackled in any quantum gravity formalism. We refer to recent literature for more details (Maudlin 2007; Huggett and Wüthrich 2018).

To summarize, the existence of new types of physical degrees of freedom, of a non-spatiotemporal type (in particular, different from continuum quantum fields), suggested by several quantum gravity approaches, points to an emergent nature of space, time, and geometry (and matter). It enlarges greatly the scope of quantum gravity research by requiring a focus on such emergence (which includes the continuum limit of the fundamentally discrete quantum gravity structures) and by raising a large number of conceptual issues. These include both ontological questions about the nature of spacetime and of its more fundamental 'constituents', and epistemological questions about their empirical significance and accessibility.

Notice that, while the technical issues related to the emergence of spacetime in such quantum gravity approaches are not much affected by the interpretation of the 'atoms of space', the conceptual issues listed above certainly are. Even if we regard them as mere technical tools encoding some sort of regularization or representing simply an intermediate step towards the true definition of the theory in terms of quantized continuum fields, the problem of the continuum limit remains the key one to tackle, via coarse graining and renormalization, as it remains necessary to devise observables that encode continuum physics in terms of the discrete building blocks one uses at first. In this case, however, no new conceptual issue arises with respect to level 0, since no meaning is assigned to any part of the theory before the same continuum limit is taken and the formulation of the theory in terms of continuum quantum fields is achieved. Not so if we give a realistic interpretation to the atoms of space suggested by the formalism and thus we investigate their physics and metaphysics even before spacetime has emerged.

2.4 Level 2: Non-Geometric Phases—The Atoms of Space(time) Are Really Not Spatiotemporal

If we take a realistic stance towards the non-spatiotemporal atoms of space, we should be ready for even more conceptual challenges.

Moving along the direction of increasing numbers of them, thus exploring their collective behaviour and continuum limit, we should expect to find that the continuum limit of such system is not unique. This is what generally (there are of course exceptions, which we treat as such) happens for any system of many interacting quantum degrees of freedom. The quantum dynamics of such interacting systems leads normally to different macroscopic phases, separated by phase transitions. Each macroscopic phase is characterized by different emergent properties, different macroscopic observables, and a different effective dynamics. In some sense, the underlying microscopic quantum system is 'replaced' by a very different kind of emergent, macroscopic system in each phase. A different macroscopic phase is, in many ways, a 'different world' (Strocchi 2015).

For our non-spatiotemporal, quantum gravity system of atoms of space, the key issue becomes then to identify such macroscopic phases and, among them, the one (or more) in which an effective description in terms of space, time, and geometry is possible, and it is governed by an effective general relativistic dynamics, at least in some approximation. In other words, the emergence of spacetime that we envisaged in the previous section should be expected to take place only in one (or some) of the possible macroscopic phases, in which the fundamental non-spatiotemporal atoms of space organize themselves. It is the task of quantum gravity formalisms that suggest fundamental non-geometric atoms of space to show that there exists such a geometric, spatiotemporal phase, in a continuum limit, in some approximation.

Quantum gravity approaches have embraced this task and have obtained considerable progress in recent years. New phases (alternative to the usually non-geometric ones in which the formalisms are first defined) have been studied in the loop quantum gravity context (Koslowski 2007; Koslowski and Sahlmann 2012; Bahr, Dittrich, and Geiller 2015; Dittrich and Geiller 2015, 2017) and in GFT (Kegeles, Oriti, and Tomlin 2018), where condensate states have also been put in correspondence with effective cosmological dynamics (Gielen, Oriti, and Sindoni 2013; Gielen and Sindoni 2016; Oriti, Sindoni, and Wilson-Ewing 2016; Oriti 2017). Indications of phase transitions have been obtained in spin foam models (Dittrich 2012; Bahr, Dittrich, Hellmann, and Kaminski 2013; Bahr and Steinhaus 2017; Dittrich 2017) and again in the GFT context (Geloun and Rivasseau 2013; Carrozza, Oriti, and Rivasseau 2014; Benedetti, Geloun, and Oriti 2015; Carrozza 2016; Carrozza, Lahoche, and Oriti 2017). Extensive studies of the phase diagram of simplicial quantum geometries, and supporting evidence for an extended De Sitter-like geometric phase, are at the core of causal dynamical triangulations (Ambjørn, Gizbert-Studnicki, Görlich, Jurkiewicz, and Németh 2018). Similar work has started recently in the causal set programme (Glaser, O'Connor, and Surya 2018). More examples could be mentioned.

One important note concerns the coupling constants, or other parameters, which characterize the quantum gravity phase diagram. In simplicial quantum gravity approaches, they are usually identified directly with the same coupling constants of continuum gravitational theories (Newton's constant, cosmological constant, etc.), since the quantum dynamics is defined as a discretization of this. In spin foam models, the situation is similar, but new parameters may enter, motivated by the specific model building guidelines or by the renormalization schemes. In GFTs, as well as in tensor models, the matching with gravitational dynamics is searched for only in a continuum approximation (even if it can be in principle performed also at the discrete level, as in simplicial quantum gravity). This matching is anyway a necessity in any quantum gravity formalism.

Our main point, now, is the following. In the presence of a non-trivial macroscopic phase diagram, thus of different macroscopic phases, and with only some of them (hopefully) of a spatiotemporal and geometric nature, spacetime and geometry can be said to be *emergent* (thus, not fundamental) in a deeper and more radical sense than already exposed at level 1.

Of the new issues raised by the existence of non-spatiotemporal atoms of space, mentioned in the previous section, the epistemological ones are not much affected. The ontological ones are.

The reason is that such atoms of space are now deprived even more of any spatiotemporal attribute, even though they remain, mathematically, the very same entities identified at level 1. In fact, whatever properties of such entities end up producing spatiotemporal observables or dynamics (e.g. some 'volume/extension attributes'), after coarse graining or some other approximation, or after being treated in a collective manner, they do so *only* in some specific phase of the system (e.g. only for specific values of the coupling constants or macroscopic parameters characterizing it). They may be 'seeds' of an emergent spacetime, in some sense, but one is precluded the possibility to consider them 'spatiotemporal properties' in disguise. Yet in other words, something more radical is at play than a simple 'approximation'. For example, continuum spacetime and geometry are not just an approximate construction from a discrete and quantum spacetime and geometry, differing only in some aspects but sharing the same nature. The very same entities, even when looked at in the same macroscopic approximation or treated by analogous coarse-graining techniques, may not produce a continuum spacetime or geometry at all. Their ontology has to be understood as being of a truly different kind. In parallel with it, the ontological status of continuum spacetime and geometry has also to be understood differently, since it turns out to be emergent in an even more radical sense.

Of course, new issues arise also at the epistemological level, at least in the sense that many of the same questions raised at level 1 have to be further refined in the presence of new quantum gravity phases for our universe. Any further analysis of such epistemological refinements, as well as of the new ontological issues raised at this new level, will have to be carried out in the context of specific quantum gravity formalism. In any case, the very existence of such new issues is the reason to emphasize the existence of such new levels of spacetime disappearance (and emergence).

2.5 Level 3: Geometrogenesis—The Emergence of Spacetime via a Phase Transition as a Physical Process

The process of deepening and broadening our perspective on spacetime disappearance and emergence, stimulated by the hypothesis of new non-spatiotemporal entities underlying the universe, proceeds even further once the possibility of new macroscopic (continuum) phases is granted. If a realistic interpretation of the fundamental atoms of space is valid, and they can organize themselves in different collective phases, there is no obvious reason why the phase transition separating non-geometric from geometric phases should not be regarded as physical as well.

This phase transition, dubbed 'geometrogenesis', and the mechanisms producing it, has been studied in a number of quantum gravity formalisms. From a more physical perspective, it was first discussed in Konopka, Markopoulou, and Smolin (2006) in a graph-based approach to quantum gravity, and immediately afterwards in the GFT formalism (Oriti 2008). More recently, it has been discussed in relation to the phase diagram of causal dynamical triangulations as well (Mielczarek 2017). Its conceptual aspects, on the other hand, have received little attention until now (see Huggett and Wüthrich 2018 for recent work).

Investigations of such conceptual aspects, however, will have to rely on a better understanding of the physical nature of the geometrogenesis phase transition as a physical process. But what type of physics does it capture?

One natural hypothesis is that it should be given a cosmological interpretation, as the process that underlies (or replaces) the big bang, as the origin of the physical universe as described by GR and quantum field theory. This is the suggestion made in the mentioned studies of geometrogenesis, and it has also been explored from a tentative phenomenological perspective in a cosmological context in Magueijo, Smolin, and Contaldi (2007). It is also the underlying hypothesis of GFT condensate cosmology (Gielen, Oriti, and Sindoni 2013; de Cesare, Pithis, and Sakellariadou 2016; Gielen and Sindoni 2016; Oriti, Sindoni, and Wilson-Ewing 2016; Oriti 2017), where geometrogenesis is technically implemented as a *condensation* of the microscopic atoms of space, with the emergent universe described in analogy with a quantum fluid.[5] It resonates as well with the so-called emergent universe scenario for primordial cosmology, an alternative to cosmic inflation, first proposed in Ellis, Murugan, and Tsagas (2003) and also realized in the context of string gas cosmology (Brandenberger 2011).[6]

While this cosmological interpretation is suggestive and, indeed, natural, it is also tricky and prone to misunderstanding. The main difficulty is the immediate temptation to interpret a cosmological phase transition not only as a physical but also as a *temporal* process. This is also a problem with the very language we use to characterize physical *processes*. A phase transition is pictured as the outcome of 'evolution' in the phase diagram of the theory, or of a 'flow' of its coupling constants; we say we 'move' towards the cosmological, geometric phase from the non-geometric, non-spatiotemporal phase, or vice versa. However, we are dealing with a system which is already described at level 2: there is no continuum space, no continuum time, no geometry in the usual sense; and it is also not characterized by features which are just 'one approximation away' from time and space.

So, first, we need to have a background-independent and non-spatiotemporal notion of 'evolution' in the space of quantum gravity coupling constants, i.e. in the 'theory space' characterizing the quantum gravity formalism at hand. Notice

[5] In the GFT cosmology context, the idea of geometrogenesis as replacing the big bang competes with the alternative idea of a bouncing scenario, as discovered in the simplest hydrodynamic description of the system.

[6] Here, however, the emergence process does not involve the temporal aspects of the universe, since a time direction remains well defined during the whole cosmic evolution, even across the phase transition.

that such evolution will relate different continuum theories, in particular different macroscopic effective dynamics, for the same fundamental quantum entities. This notion of evolution in theory space is what specific renormalization group (RG) schemes in various quantum gravity formalism will provide.

Next, we can ask whether any notion of 'proto-time' and 'proto-temporal' evolution can be associated with such flow in a quantum gravity theory space, and how it relates to any of the notions of time that may emerge in the geometric, spatiotemporal phase of the universe (thus, 'after' the geometrogenesis phase transition).

There are two orders of difficulties here. One is the mentioned absence of any notion of time at level 2, which adds conceptual difficulties to the absence of any notion of time of level 1, and to the 'problem of time' in (quantum) GR, i.e. level 0. The other is that, strictly speaking, even the standard RG flow of coupling constants in ordinary statistical (field) theory is 'timeless' and not interpreted as standard evolution, since it may well refer to systems *at equilibrium*.[7] The reason why we have no particular conceptual issue in understanding the flow in theory space and the approach to phase transitions in temporal terms, despite the fact that they refer to a change in the time-independent coupling constants of the system, is that we can easily imagine an external observer (the experimental physicist in the lab) tuning such coupling constants towards their critical values, and thus pushing the system towards the relevant phase transition. Needless to say, no such external observer is available in quantum gravity.

Any notion of time or, better, 'proto-time' that could be associated to such flow across the quantum gravity phase diagram would in any case deserve such name only in the sense that, once used to parametrize the flow across a non-geometric phase towards a geometrogenesis phase transition, it ends up matching some spatiotemporal observable that can be used as a time variable within the geometric phase. Vice versa, it would correspond to what is left of some geometric variable used to define a notion of time in such phase, and used as well as a notion of RG scale for the quantum gravity system, once the same system flows across a geometrogenesis phase transition into a non-spatiotemporal phase.

We leave a detailed and concrete analysis of this problem, and of the many conceptual issues associated to it, to future work. Here, we only suggest a possible strategy, which can be understood as 'pushing the relational framework (used to obtain a notion of time at level 0) two levels forward'. The idea would be to take an internal (dynamical) degree of freedom, used as a relational clock in the geometric GR-governed phase at level 0, to parametrize (as the relevant notion of 'scale') the RG flow of the underlying non-spatiotemporal quantum gravity system.

[7] Of course, when one is dealing with systems out of equilibrium, and thus time-dependent, this additional difficulty is absent.

This would allow us to give a proto-temporal evolution interpretation to the same RG flow, even in the non-geometric phases, and thus to the geometrogenesis phase transition. One example could be the emergent (free, massless) scalar field used in GFT cosmology (de Cesare, Pithis, and Sakellariadou 2016; Oriti, Sindoni, and Wilson-Ewing 2016; Gielen and Oriti 2018) as well as in (loop) quantum cosmology (Agullo and Singh 2017). Another could be any observable playing the role of 'volume' or scale factor of the universe, in the geometric phase; one could imagine then the RG flow to drive the system and its coupling constants from large volumes towards very small volumes and there hitting a geometrogenesis phase transition (thus replacing the big bang singularity), ushering the universe into the non-spatiotemporal phase (where the interpretation of the same variable as 'volume' or relational time would cease, contextually, to make sense).

A scenario of the type sketched above, we believe, will require serious reflections not only on the nature of space and time, but also on the RG when applied to quantum gravity and on a covariant, spacetime-free understanding of statistical mechanics. Even more clearly, it may have profound implications on the philosophy of cosmology that will be subject to a broadening of scope and perspective in parallel with the one we are suggesting for the philosophy of quantum gravity.

In the end, these many new conceptual issues that arise in this scenario are the reason to associate to it a new level of spacetime disappearance and emergence.

2.6 An Analogy: From the Atoms to the Hydrodynamics of (Super)fluids

Before concluding, we would like to offer a physical analogy of the situation outlined for quantum gravity, and of the various levels of 'emergence' we illustrated in this contribution. We hope this will clarify further the conceptual framework we have in mind. For more details on this example, see Volovik (2008).

Consider the hydrodynamic description of a fluid, with the main dynamical variables being the fluid density and velocity, and interesting observables being the total momentum and energy, vorticity, circulation of vortex excitations, viscosity, etc., which are functions of them. On top of the global configurations of the fluid, one has propagating excitations over them corresponding to sound waves with their own characteristic dispersion relation. Notice that one can also consider extended versions of standard fluid hydrodynamics, including additional functions of the same density and velocity fields (e.g. gradient terms); this is the case, for example, of superfluid hydrodynamics. In our analogy, standard hydrodynamics would be the counterpart of GR, with spatiotemporal, geometric observables (volumes, areas, distances, time intervals, curvature, etc.) corresponding to various hydrodynamic observables, which are functions of the basic fields (in GR, the metric or metric and connection etc., plus matter fields, depending on the specific formulation of classical gravity one uses).

From this classical theory, one can move to the quantum regime. One could start by simply quantizing the classical theory. The resulting quantum theory is perturbatively non-renormalizable as a quantization of sound waves (in fact, with the same order of divergences as perturbative GR in terms of gravitons) but it could make sense non-perturbatively. In the presence of gauge symmetries, these will carry over to the quantum level and require observables to satisfy them (the analogue of the problem of time in quantum gravity). Emergent notions of time and space associated to observables constructed via the relational strategy correspond, in the analogy, to specific hydrodynamic observables, also at the quantum level but thus subject to extra complications due to their quantum nature. The same story goes for superfluid hydrodynamics. This is our level 0.

We know that there is a deeper level of description (and of physics) for fluids, which becomes even more relevant for superfluids: the underlying quantum theory of atoms. This could be described by a quantum field theory for bosonic or fermionic entities, and we know that its features are not captured by the straightforward quantization of classical hydrodynamics. The microscopic quantum dynamics differs from the one corresponding to quantum hydrodynamics (even when expressed in the same variables, as in the case of superfluids described in terms of a single complex scalar field); moreover, the classical, long-wavelength approximation of quantum hydrodynamics does not reproduce the classical theory one started from, in general. For superfluids, moreover, we know that the quantum statistics of the atoms is required in a complete understanding of the key macroscopic features, i.e. superfluidity itself. The macroscopic parameters are understood as functions of the microscopic coupling constants, e.g. the mass of the atoms. These are instances of the fact that a continuum approximation built on many fundamental degrees of freedom is conceptually and physically different from the simple classical approximation of the same degrees of freedom, a point emphasized in Oriti (2020) in the case of quantum gravity. The very basic variables of the macroscopic theory, the hydrodynamic variables, are also understood to be useful expressions of collective (and coarse-grained) properties of the fundamental atoms, and the same holds of course at the level of observables. In particular, it holds for those observables that, in the analogy, correspond to spatiotemporal and geometric quantities. Spacetime is emergent, then, in the same sense in which the hydrodynamic description is emergent from the more fundamental atomic description, based on a different kind of degrees of freedom, thus a different ontology. This is level 1.

Broadening further the picture of the fundamental atomic system, we know that the fluid description in terms of hydrodynamics and, more specifically, of superfluid hydrodynamics is only available in one of its macroscopic, continuum phases, the one in which it behaves indeed as a liquid. The same system can organize itself also as a gas or a solid (more exotic condensed matter systems may have an even richer phase diagram). In these other phases, the observables that correspond, in our analogy, with spatiotemporal or geometric observables will simply not be available, in general. The very hydrodynamic variables used to construct them, and analogous to the spatiotemporal continuum fields of GR, will not be available. Notice that this is an additional, more radical point than the noted difference between the two hydrodynamic descriptions of ordinary fluids and superfluids, i.e. the two different effective (and emergent) dynamics of the same atomic system in two different fluid phases. The latter may correspond, in the analogy, to two different gravitational theories or effective spatiotemporal dynamics in two different geometric phases of a fundamental non-spatiotemporal system of 'atoms of space', with physically distinguished properties. If it was already clear that atoms were not simply 'pieces of fluids' or 'fluid-like' in any sense at level 1, now that we notice they can form solids or gases or other non-fluid phases, we are truly forced to think of them as possessing a different, independent nature. This is level 2.

There is also no doubt, in the case of ordinary fluids and superfluids, that the phase transitions separating different macroscopic phases, and in particular the fluid one(s) and the non-fluid ones, correspond to physical processes. The analogues of our additional questions concerning spacetime in the presence of physical phase transitions of our fundamental quantum gravity system would go roughly as follows. We would ask what happens to specific hydrodynamic observables across the phase transitions and whether there is any remnant of them, or simply a corresponding observable, constructed in the same way out of the atomic degrees of freedom, in the non-fluid phases. We would then try to use them to characterize the physics of the phase transition itself. We could also ask whether the same observables illuminate aspects of the RG flow across the phase diagram, i.e. the running of atomic coupling constants, and in particular if they could be used as the relevant 'scale' parametrizing the same flow. In any case, when studying the meaning and physics of phase transitions of the non-fluid-like atomic system, while keeping our attention on some hydrodynamic observables, those corresponding in the analogy to spatiotemporal and geometric quantities, we would be working at level 3.

2.7 Conclusions

We have explored the issue of spacetime emergence in quantum gravity, by articulating several levels at which this emergence (correspondingly, the disappearance of space and time) can be intended. As we have shown, these levels correspond to successive steps in a process of widening of the perspective on the nature of space, time, and geometry, revealing new questions at each step. We have highlighted some of the new technical issues that arise at each level, and discussed in particular the new conceptual issues. The main goal was to clarify the broader perspective one can take towards the issue of the emergence of spacetime in quantum gravity, forming the basis for a lot of recent work in the field. We believe that this broader perspective deepens the scope of the debate on the nature of spacetime, both philosophically and physically. Finally, we hope that it could be the starting point for many new research directions in the philosophy of quantum gravity and of spacetime, for what concerns both its metaphysical and epistemological aspects.

References

Agullo, I. and P. Singh (2017). Loop quantum cosmology. In A. Ashtekar and J. Pullin (Eds.), *Loop Quantum Gravity: The First 30 Years*, pp. 183–240. Singapore: World Scientific.

Albert, D. Z. (2000). *Time and Chance*. Cambridge, MA: Harvard University Press.

Ambjørn, J., J. Gizbert-Studnicki, A. Görlich, J. Jurkiewicz, and D. Németh (2018). The phase structure of causal dynamical triangulations with toroidal spatial topology. *Journal of High Energy Physics 2018*(6), 111.

Ambjørn, J., A. Görlich, J. Jurkiewicz, and R. Loll (2012). Nonperturbative quantum gravity. *Physics Reports 519*(4–5), 127–210.

Ashtekar, A. and J. Lewandowski (2004). Background independent quantum gravity: A status report. *Classical and Quantum Gravity 21*(15), R53–R152.

Bahr, B., B. Dittrich, and M. Geiller (2015). A new realization of quantum geometry. arXiv 1506.08571.

Bahr, B., B. Dittrich, F. Hellmann, and W. Kaminski (2013). Holonomy spin foam models: Definition and coarse graining. *Physical Review D 87*(4), 044048.

Bahr, B. and S. Steinhaus (2017). Hypercuboidal renormalization in spin foam quantum gravity. *Physical Review D 95*(12), 126006.

Bain, J. (2013). The emergence of spacetime in condensed matter approaches to quantum gravity. *Studies in History and Philosophy of Modern Physics 44*(3), 338–345.

Batterman, R. (2006). Reduction and renormalization. In *The Robert and Sarah Boote Conference in Reductionism and Anti-reductionism in Physics*, Pittsburgh, PA, April 22–23. http://philsci-archive.pitt.edu/2852.

Batterman, R. W. (2011). Emergence, singularities, and symmetry breaking. *Foundations of Physics 41*(6), 1031–1050.

Bedau, M. and P. Humphreys (Eds.) (2008). *Emergence*. Cambridge, MA: MIT Press.

Benedetti, D., J. B. Geloun, and D. Oriti (2015). Functional renormalisation group approach for tensorial group field theory: A rank-3 model. *Journal of High Energy Physics 03*, 84.

Blau, M. and S. Theisen (2009). String theory as a theory of quantum gravity: A status report. *General Relativity and Gravitation 41*(4), 743–755.

Bodendorfer, N. (2016). An elementary introduction to loop quantum gravity. arXiv 1607.05129.

Brandenberger, R. H. (2011). String gas cosmology: Progress and problems. *Classical and Quantum Gravity 28*(20), 204005.

Butterfield, J. (2011a). Emergence, reduction and supervenience: A varied landscape. *Foundations of Physics 41*(6), 920–959.

Butterfield, J. (2011b). Less is different: Emergence and reduction reconciled. *Foundations of Physics 41*(6), 1065–1135.

Butterfield, J. and N. Bouatta (2012). Emergence and reduction combined in phase transitions. In J. Kouneiher, C. Barbachoux, and D. Vey (Eds.), *Proceedings of Frontiers of Fundamental Physics (AIP)*, Volume 11, pp. 383–403. Melville, NY: AIP.

Butterfield, J. and C. J. Isham (2001). Spacetime and the philosophical challenge of quantum gravity. In C. Callender and N. Huggett (Eds.), *Physics Meets Philosophy at the Planck Scale*, pp. 33–89. Cambridge: Cambridge University Press.

Carrozza, S. (2016). Flowing in group field theory space: A review. *SIGMA 12*, 070.

Carrozza, S., V. Lahoche, and D. Oriti (2017). Renormalizable group field theory beyond melonic diagrams: An example in rank four. *Physical Review D 96*(6), 066007.

Carrozza, S., D. Oriti, and V. Rivasseau (2014). Renormalization of a SU(2) tensorial group field theory in three dimensions. *Communications in Mathematical Physics 330*(2), 581–637.

Crowther, K. (2014). *Appearing Out of Nowhere: The Emergence of Spacetime in Quantum Gravity*. Ph.D. thesis, University of Sydney, arXiv 1410.0345.

Dainton, B. (2001). *Time and Space*. Montreal: McGill-Queen's University Press.

de Cesare, M., A. G. Pithis, and M. Sakellariadou (2016). Cosmological implications of interacting group field theory models: Cyclic universe and accelerated expansion. *Physical Review D 94*(6), 064051.

Delcamp, C. and B. Dittrich (2017). Towards a phase diagram for spin foams. *Classical and Quantum Gravity 34*(22), 225006.

Dittrich, B. (2012). From the discrete to the continuous: Towards a cylindrically consistent dynamics. *New Journal of Physics 14*(12), 123004.

Dittrich, B. (2017). The continuum limit of loop quantum gravity: A framework for solving the theory. In A. Ashtekar and J. Pullin (Eds.), *Loop Quantum Gravity: The First 30 Years*, pp. 153–179. Singapore: World Scientific.

Dittrich, B. and M. Geiller (2015). A new vacuum for loop quantum gravity. *Classical and Quantum Gravity 32*(11), 112001.

Dittrich, B. and M. Geiller (2017). Quantum gravity kinematics from extended TQFTs. *New Journal of Physics 19*(1), 013003.

Dowker, F. (2013). Introduction to causal sets and their phenomenology. *General Relativity and Gravitation 45*(9), 1651–1667.

Earman, J. (1989). *World Enough and Spacetime*. Cambridge, MA: MIT Press.

Ellis, G., J. Murugan, and C. Tsagas (2003). The emergent universe: An explicit construction. *Classical and Quantum Gravity 21*(1), 233–250.

Gambini, R. and R. A. Porto (2001). Relational time in generally covariant quantum systems: Four models. *Physical Review D 63*(10), 105014.

Geloun, J. B., R. Martini, and D. Oriti (2016). Functional renormalization group analysis of tensorial group field theories on $\mathbb{R}d$. *Physical Review D 94*(2), 024017.

Geloun, J. B. and V. Rivasseau (2013). A renormalizable 4-dimensional tensor field theory. *Communications in Mathematical Physics 318*(1), 69–109.

Gielen, S. and D. Oriti (2018). Cosmological perturbations from full quantum gravity. *Physical Review D 98*(10), 106019.

Gielen, S., D. Oriti, and L. Sindoni (2013). Cosmology from group field theory formalism for quantum gravity. *Physical Review Letters 111*(3), 031301.

Gielen, S. and L. Sindoni (2016). Quantum cosmology from group field theory condensates: A review. *SIGMA 12*, 082.

Glaser, L., D. O'Connor, and S. Surya (2018). Finite size scaling in 2d causal set quantum gravity. *Classical and Quantum Gravity 35*(4), 045006.

Gurau, R. and J. P. Ryan (2012). Colored tensor models: A review. *SIGMA 8*, 020.

Hamber, H. W. (2009). Quantum gravity on the lattice. *General Relativity and Gravitation 41*(4), 817–876.

Hohm, O., D. Lüst, and B. Zwiebach (2013). The spacetime of double field theory: Review, remarks, and outlook. *Fortschritte der Physik 61*(10), 926–966.

Huggett, N. and C. Wüthrich (2013). Emergent spacetime and empirical (in)coherence. *Studies in History and Philosophy of Modern Physics 44*(3), 276–285.

Huggett, N. and C. Wüthrich (2018). The (a)temporal emergence of spacetime. *Philosophy of Science 85*(5), 1190–1203.

Isham, C. J. (1991). Conceptual and geometrical problems in quantum gravity. In H. Mitter and G. H. (Eds.), *Recent Aspects of Quantum Fields*, pp. 123–229. Berlin, Heidelberg: Springer.

Kegeles, A., D. Oriti, and C. Tomlin (2018). Inequivalent coherent state representations in group field theory. *Classical and Quantum Gravity 35*(12), 125011.

Kiefer, C. (2006). Quantum gravity: General introduction and recent developments. *Annalen Der Physik 15*(1–2), 129–148.

Kiefer, C. (2013). Conceptual problems in quantum gravity and quantum cosmology. *ISRN Mathematical Physics 509316*. arXiv 1401.3578.

Konopka, T., F. Markopoulou, and L. Smolin (2006). Quantum graphity. arXiv hep-th/0611197.

Koslowski, T. A. (2007). Dynamical quantum geometry (DQG programme). arXiv 0709.3465.

Koslowski, T. and H. Sahlmann (2012). Loop quantum gravity vacuum with nondegenerate geometry. *SIGMA 8*, 026.

Krajewski, T. (2013). Group field theories. *Proceedings of Science 140 (3rd Quantum Gravity and Quantum Geometry School (QGQGS 2011))*, 005.

Lam, V. and M. Esfeld (2013). A dilemma for the emergence of spacetime in canonical quantum gravity. *Studies in History and Philosophy of Modern Physics 44*(3), 286–293.

Lam, V. and C. Wüthrich (2018). Spacetime is as spacetime does. *Studies in History and Philosophy of Modern Physics 64*, 39–51.

Magueijo, J., L. Smolin, and C. R. Contaldi (2007). Holography and the scale invariance of density fluctuations. *Classical and Quantum Gravity 24*(14), 3691–3700.

Maudlin, T. (2007). Completeness, supervenience and ontology. *Journal of Physics A: Mathematical and Theoretical 40*(12), 3151.

Mielczarek, J. (2017). Big bang as a critical point. *Advances in High Energy Physics*, 4015145.

Mielczarek, J. and T. Trześniewski (2018). Towards the map of quantum gravity. *General Relativity and Gravitation 50*(6), 68.

Norton, J. (2003). General covariance, gauge theories, and the Kretschmann objection. In K. Brading and E. Castellani (Eds.), *Symmetries in Physics: Philosophical Reflections*. Cambridge: Cambridge University Press.

Oriti, D. (2008). Group field theory as the microscopic description of the quantum spacetime fluid: A new perspective on the continuum in quantum gravity. *Proceedings of Science 043 (From Quantum to Emergent Gravity: Theory and Phenomenology (QG-Ph))*, 030.

Oriti, D. (2009). *Approaches to Quantum Gravity: Toward a New Understanding of Space, Time and Matter*. Cambridge: Cambridge University Press.

Oriti, D. (2012). The microscopic dynamics of quantum space as a group field theory. In G. Ellis, J. Murugan, and A. Weltman (Eds.), *Foundations of Space and Time: Reflections on Quantum Gravity*, pp. 257–320. Cambridge: Cambridge University Press.

Oriti, D. (2014). Disappearance and emergence of space and time in quantum gravity. *Studies in History and Philosophy of Modern Physics 46*, 186–199.

Oriti, D. (2017). The universe as a quantum gravity condensate. *Comptes Rendus Physique 18*(3–4), 235–245.

Oriti, D. (2018). Spacetime as a quantum many-body system. In G. Angilella and C. Amovilli (Eds.), *Many-body Approaches at Different Scales*, pp. 365–379. Cham: Springer.

Oriti, D. (2019). No alternative to proliferation. In R. Dardashti, R. Dawid, and K. Thébault (Eds.), *Why Trust a Theory?*, pp. 125–153. Cambridge: Cambridge University Press.

Oriti, D. (2020). The Bronstein hypercube of quantum gravity. In N. Huggett, K. Matsubara, and C. Wüthrich (Eds.), *Beyond Spacetime: The Philosophical Foundations of Quantum Gravity*, pp. 25–52. Cambridge: Cambridge University Press.

Oriti, D., L. Sindoni, and E. Wilson-Ewing (2016). Emergent Friedmann dynamics with a quantum bounce from quantum gravity condensates. *Classical and Quantum Gravity 33*(22), 224001.

Padmanabhan, T. (2015). Emergent gravity paradigm: Recent progress. *Modern Physics Letters A 30*(03n04), 1540007.

Page, D. N. and W. K. Wootters (1983). Evolution without evolution: Dynamics described by stationary observables. *Physical Review D 27*(12), 2885.

Perez, A. (2013). The spin-foam approach to quantum gravity. *Living Reviews in Relativity 16*(3).

Pooley, O. (2010). Substantive general covariance: Another decade of dispute. In M. Suárez, M. Dorato, and M. Rédei (Eds.), *EPSA Philosophical Issues in the Sciences*, pp. 197–209. Dordrecht: Springer.

Rivasseau, V. (2016). Random tensors and quantum gravity. *SIGMA 12*, 069.

Rovelli, C. (1991a). Time in quantum gravity: An hypothesis. *Physical Review D 43*(2), 442–456.

Rovelli, C. (1991b). What is observable in classical and quantum gravity? *Classical and Quantum Gravity 8*(2), 297–316.

Rovelli, C. (2020). Space and time in loop quantum gravity. In N. Huggett, K. Matsubara, and C. Wüthrich (Eds.), *Beyond Spacetime: The Philosophical Foundations of Quantum Gravity*, pp. 117–132. Cambridge: Cambridge University Press.

Rovelli, C. and M. Gaul (2000). Loop quantum gravity and the meaning of diffeomorphism invariance. In J. Kowalski-Glikman (Ed.), *Towards Quantum Gravity*, pp. 277–324. Berlin: Springer.

Seiberg, N. (2007). Emergent spacetime. In D. Gross, M. Henneaux, and A. Sevrin (Eds.), *The Quantum Structure of Space and Time*, pp. 163–213. Singapore: World Scientific.

Strevens, M. (2008). *Depth: An Account of Scientific Explanation*. Cambridge, MA: Harvard University Press.

Strocchi, F. (2015). *Symmetry Breaking*. Berlin: Springer.

Teitelboim, C. (1982). Quantum mechanics of the gravitational field. *Physical Review D 25*, 3159.

Thiemann, T. (2001). Introduction to modern canonical quantum general relativity. arXiv gr-qc/0110034v1.

van Inwagen, P. and M. Sullivan (2018). Metaphysics. In E. N. Zalta (Ed.), *The Stanford Encyclopedia of Philosophy* (Spring 2018 ed.). Metaphysics Research Lab, Stanford University.

Volovik, G. (2008). From quantum hydrodynamics to quantum gravity. In H. Kleiner and R. B. Jantzen (Eds.), *The Eleventh Marcel Grossmann Meeting: On Recent Developments in Theoretical and Experimental General Relativity, Gravitation and Relativistic Field Theories*, pp. 1451–1470. Singapore: World Scientific.

Wüthrich, C. (2019). The emergence of space and time. In S. Gibb, R. Hendry, and T. Lancaster (Eds.), *The Routledge Handbook of Emergence*, pp. 315–326. New York: Routledge.

Wüthrich, C. (2020). When the actual world is not even possible. In G. Darby, D. Glick, and A. Marmodoro (Eds.), *The Foundation of Reality: Fundamentality, Space and Time*, pp. 233–253. Oxford: Oxford University Press.

3

On Dualities and Equivalences between Physical Theories

Jeremy Butterfield

Abstract

The main aim of this paper is to make a remark about the relation between (i) dualities between theories, as 'duality' is understood in physics and (ii) equivalence of theories, as 'equivalence' is understood in logic and philosophy. The remark is that in physics, two theories can be dual, and accordingly get called 'the same theory', though we interpret them as disagreeing—so that they are certainly *not* equivalent, as 'equivalent' is normally understood. So the remark is simple, but, I shall argue, worth stressing, since often neglected. My argument for this is based on the account of duality developed by De Haro. I also spell out how this remark implies a limitation of proposals (both traditional and recent) to understand theoretical equivalence as either logical equivalence or a weakening of it.

Dedicated to Graeme Segal

Motto: "I defy anyone to avoid getting confused by active vs. passive transformations". (Graeme Segal, in conversation about Einstein's hole argument, 2006)

3.1 Introduction

3.1.1 Prospectus: The Remark and the Implication

In recent philosophy of physics, there has been a surge of interest in two topics:

(i) how best to understand dualities between physical theories;
(ii) how best to individuate physical theories, and so how best to define theories being equivalent.

Jeremy Butterfield, *On Dualities and Equivalences between Physical Theories* In: *Philosophy Beyond Spacetime: Implications from Quantum Gravity*. Edited by: Christian Wüthrich, Baptiste Le Bihan, and Nick Huggett, Oxford University Press. DOI: 10.1093/oso/9780198844143.003.0003

These topics are closely related, since all agree that a duality is a matter of two theories being in some sense 'the same'. And since dualities have been, for some twenty-five years, centre-stage in the dominant research programme for quantum gravity, viz. string theory, the philosophical examination of dualities, and so of theoretical equivalence, is indeed a topic for 'philosophy beyond spacetime'.

This chapter proposes, however, to take a step back from dualities in string theory, which is, after all a speculative subject, most of whose dualities are, so far, conjectures rather than theorems. Indeed, my main aim is cautionary: to stress that in physics, two theories can be dual, and accordingly get called 'the same theory', though we interpret them as disagreeing—so that they are certainly *not* equivalent, as 'equivalent' is normally understood, both in everyday language and in philosophy.

I will establish this point—I shall say *Remark*—by: (a) reporting the account of duality in physics, developed by De Haro (2016, 2020), which I endorse and which we (2017) call a 'Schema' (sections 3.2 and 3.3); and then (b) supporting the Schema with examples that illustrate the Remark (section 3.4). For brevity and simplicity, I give two elementary examples, one classical and one quantum. But with an eye on 'philosophy beyond spacetime', I stress at the outset that string theory's dualities also give examples (as do other parts of advanced physics).[1]

As we shall see in these examples, two dual theories can 'disagree' in either of two ways:

- (Contr): by the theories making contrary assertions about a common subject-matter (they may well agree on some claims, but they contradict each other over other claims); or
- (Diff): by the theories describing different subject-matters (though the descriptions are 'isomorphic' or 'matching'—hence the duality).

In section 3.3, I will make precise the notion of a subject-matter. I also agree that (Diff) hardly merits the label 'disagreement', since both theories could be true. But 'disagree' is just my convenient umbrella term. The main point is the same, for either (Contr) or (Diff): in both situations, one would not say that the dual theories are 'equivalent', on any normal understanding.

After establishing the Remark in section 3.4, I spell out an *Implication*: namely, a limitation of proposals (both traditional and recent) to understand theoretical

[1] I discuss two string theory examples, gauge/gravity duality and T-duality, in section 6 of a longer version of this chapter, available at: http://philsci-archive.pitt.edu/14736. I argue that in both examples, the two dual theories can be interpreted as disagreeing, in one of the senses that I label (Contr) or (Diff). For gauge/gravity duality, the disagreement is about, in the first instance, the dimensionality of space, while for T-duality, it is about the radius of a compact dimension of space.

equivalence either as logical equivalence or as a weakening of it (section 3.5). In short: two disagreeing dual theories might be formalized so as to be logically equivalent. And this implies that logical equivalence is too weak an explication of theoretical equivalence—as is, therefore, any of the recently proposed weakenings of logical equivalence.

Thus the chapter has two main stages. The first stage, sections 3.2 to 3.4, establishes the Remark, with elementary examples from physics. The 'tone' is philosophy of physics, rather than of logic. But in the second stage, section 3.5, philosophy of logic comes to the fore. Indeed, readers interested more in logic than in physics can read section 3.5 without reading the details of sections 3.2 to 3.4.

As we shall see, the Remark is, in essence, simple—'elementary, my dear Watson'—in two ways. First: it is illustrated by some familiar, indeed traditional, cases, which are examples of our Schema for duality. My example of (Contr) will be different (i.e. contrary) specifications of absolute rest in Newtonian mechanics, set in Newtonian spacetime. My example of (Diff) will be a 'reworking' of position-momentum duality in quantum mechanics. These will be Examples (1) and (2) in section 3.4. Second: the Remark, and also the Implication, turn on a familiar idea: that formal mathematical methods can only 'discern structure', and so cannot 'cut finer' than isomorphism, i.e. cannot distinguish isomorphic copies. The classification of an Example as a case of (Contr) (or of Diff) therefore goes beyond formal matters such as isomorphism, and depends on interpretation.

This appeal to interpretation prompts two more general, but again familiar, morals of what follows.

The first moral is about the need for judgement—and the difficulty of distinguishing 'fact' from 'convention'. In logic class, we learn to respect the gap between formal logical systems and arguments in natural language; and to admit that regimenting such an argument in a system requires some judgement-calls. Similarly here: aligning physical theories with our formal definition of duality will require some judgements. These judgements will impinge on whether we classify an Example as a case of (Contr); and the same goes for (Diff). I am steadfast that I have classified my Examples correctly, by respecting the intended meanings of the theories' words. In section 3.3, I will make 'intended meaning' more precise (viz. by endorsing the framework of intensional semantics). But I admit to relying on intuition—or perhaps better: stipulation—about e.g. what a proponent of Newtonian mechanics means by their word 'absolute rest', or what a proponent of statistical mechanics means by 'temperature'. But—as we will see—I think this is the best one can do, thanks to lessons philosophers learnt some sixty years ago, from the troubles that beset positivists' attempts to sharply distinguish, within a scientific theory, 'fact' from 'convention'—its empirical claims from its conventional elements—once and for all.[2]

The second moral is broader. It is that, quite apart from the topic of dualities (and in particular, De Haro's and my account of them), a verdict that two theories are equivalent depends on interpretation. 'Theoretical equivalence' is of course a term of art in the post-positivist tradition, which takes 'scientific theory' as a unit of analysis and accepts some sort of theory/observation distinction. In principle, you could define it however you wish. But all hands agree that we need some notion of 'theories making the very same claims about the world, i.e. not only observational claims but also theoretical ones'. For one thing, we need some such strengthening of the idea of observational (also known as: empirical) equivalence, in order to articulate and then assess the under-determination of theory by data, namely as a matter of observationally equivalent, but theoretically inequivalent, theories. Here, the phrase 'claims about the world' obviously invokes interpretation. But interpretation is liable to be vague, and even controversial, not least because of widespread post-Quinean suspicion of meanings. Hence the long tradition of using formal ideas from logic to try and give a precise explication of theoretical equivalence. The best-known proposal takes theories in the traditional way as sets of sentences, and then defines *logical equivalence* of theories, exactly as in a logic book.

But this cannot be right as an explication of theoretical equivalence, for three reasons, each of which leads us back to interpretation. The first two are well known, and the third is the main idea of this paper's Implication.

(1) Many philosophers reject the traditional construal of theories as sets of sentences, usually by advocating the rival 'semantic conception of theories' (a view I will touch on again in section 3.5.2).

(2) For two theories to be logically equivalent, in the precise sense of a logic book, they must be formulated in the very same vocabulary. And this is certainly too restrictive an explication of 'making the very same claims about the world', since two theories could do that while using different vocabularies. Just think of classical electromagnetism written in English and French. This second point has prompted the effort, in both logic and philosophy, to give weaker definitions: that is, to define equivalence of theories more weakly than (but as precisely as) the familiar logical equivalence. (I review this effort in section 3.5.3.)

(3) As I announced above, my Remark has an Implication that 'goes in the opposite direction' from this effort. For the Remark shows that logical

[2] Here it is usual to cite Quine (1951). But I maintain that Quine was unfair to Carnap, and that there are more persuasive rebuttals of making this a sharp distinction; cf. Putnam (1962); Stein (1992).

> equivalence is in some cases too weak an explication of 'making the very same claims about the world'. Namely: dual pairs satisfying (Contr) or (Diff), once they are formalized (i.e. stated in a formal language), may well be logically equivalent, and thereby equivalent in any weaker sense—though, thanks to (Contr) or (Diff), they certainly do *not* 'make the very same claims about the world'.

Here, reasons (2) and (3) obviously form a complementary pair. (2) says that we need to allow for *synonymy*, like 'electric' in English and 'électrique' in French; and (3) says that we need to allow for *homonymy*, like 'bank' in finance and beside a river. As I said in the first moral above: I am steadfast that, though we need to exercise judgement, in my Examples the facts about synonymy and homonymy are sufficiently objective to make my Remark and Implication true.[3]

But though the Remark and Implication are simple, and these two general morals familiar, they are worth stressing, for two reasons. First: simple though they are, most of the recent philosophical literature on dualities and theoretical equivalence *fails* to state them, which can be confusing. As we shall see, the confusion is closely related to confusions that often occur when thinking about active vs. passive transformations: hence the chapter's motto. Second: they provide the mental exercise of navigating the various conflicting usages of such words as 'theory', 'model', 'equivalence', and even 'isomorphism'.

3.1.2 Relations to Other Work

To help introduce the chapter, I comment on relations to other work. The first comment is that this chapter is a companion to De Haro (2016, 2020): not just because these papers state the Schema for duality (in their first sections), but also because they emphasize that dual theories can disagree (in my umbrella sense). In De Haro's jargon, dual theories are by definition *theoretically equivalent*: so note that

[3] Of course, I am not alone in stressing that verdicts of theoretical equivalence depend on interpretation. Coffey (2014) argues for this (partly by citing philosophers disagreeing about putative examples: his sections 2, 3). I agree with him in spirit (though not in letter—I disagree with his treatment of examples, and I do not share his post-Quinean anxiety about meanings (835, 838)). A closer kindred spirit is van Fraassen, especially in his reply (2014) to Halvorson's critique (2012) of the semantic view of theories. Van Fraassen's section 2 stresses the indispensability of interpretation. In particular, he says: "the identity of a theory [cannot] be defined in terms of the corresponding set of mathematical structures without reference to their representational function" (278). And more specifically, van Fraassen prefigures my reason (3) about homonymy: "if the same diffusion equation is presented to describe gas diffusion, and, elsewhere, temperature distribution over time, would anyone think that one and only one theory was being presented?" (279). He also says the point is familiar: "these sorts of examples had ample discussion in familiar literature" (279). As I say in the main text: I agree—but maintain that nowadays, it is still worth stressing, especially in connection with dualities.

for him, 'theoretical equivalence' does *not* mean 'making the same claims about the world'. Yet, he says, dual theories can make different, i.e. disagreeing, claims about the world—which he dubs their being *physically inequivalent*. I agree with De Haro, though I shall keep to my jargon, including my colloquial umbrella word 'disagree', and (Contr) and (Diff) (section 3.5.1 gives my rationale about jargon).

As to other recent literature about dualities: here, I will just expand on the first of my reasons, at the end of section 3.1.1, for stressing the Remark. The reason why most of the philosophical literature about dualities omits the Remark (and so its Implication) is that it concentrates on dualities in string theories; and string theories are usually presented as 'theories of everything' (TOEs), 'toy cosmologies', i.e. theories of an entire possible world (although perhaps not as 'final theories', whatever that might mean!). Thus several authors hold that dualities for TOEs ('toy cosmologies') *must* be taken as making the same claims about the world: as 'agreeing', in my jargon—though often called by them, as by De Haro, 'physically equivalent'. Of course, the details vary. But there is a common intuitive idea: roughly speaking, that for two dual TOEs, there is literally 'no room in the cosmos' for them to have separate but isomorphic subject-matters. Here are three examples:

(a) Rickles says "dual descriptions are not in competition for 'physical reality': rather they are to be considered as complementary descriptions of one and the same physical situation" (2011, 55), and "theories are said to be dual when they generate the same physics" (2011, 56). And again in later work: "we can take both descriptions to be describing the same physical situation using different concepts" (Rickles 2013, 319); "there is no *competition* between dual descriptions, and the incompatibility can be transformed away in a certain sense . . . dualities are examples of equivalent descriptions *simpliciter*" (Rickles 2017, 63 and *passim*).
(b) Matsubara (2013, section 4.1, 477f) requires dual theories to be physically equivalent, viz. by requiring any differences between them not to be matters of empirical interpretation, or choices of coordinates or gauge (or at least not obviously so).
(c) Huggett (2017, 86) says "the case of dual total theories is clearly one in which the putative differences are 'hidden' in a very strong sense . . . [so that] at least from a practical, scientific point of view, it makes sense to treat those differences as non-physical . . . In other words, long established, well-motivated scientific reasoning should lead us to think that dual total theories represent the same physical situation".[4]

[4] Agreed: these authors also note that in some dualities, away from string theory or TOEs, the duals have different subject-matters. For example, Matsubara (2013, section 4.2.1, 478) and Huggett (2017, 86) note this for an elementary harmonic oscillator duality.

So to sum up this chapter's relation to these authors: I reject this consensus. Two dual theories can disagree (even in string theory, though I cannot here consider such examples—cf. footnote 1). Or more modestly, with an eye on my Newtonian mechanics example in section 3.4.1: this view of the duals is tenable—as tenable as was Newton's view (or Clarke's view, in his correspondence with Leibniz) that two different specifications of absolute rest contradict each other.

I of course agree that even for TOEs, the literature is not unanimous. A laudable exception to the consensus—a close cousin of my own position—is Read (2016), and Read and Möller-Nielsen (2020). The first of these is about string-theoretic dualities. It gives a nuanced roster of interpretative options, and seems to favour a combination of them (224f, 232), rather than saying that duals always agree, i.e. are physically equivalent (his options 2 and 3, explored on pp. 227–230). The second paper relates our topic—do some duals disagree?—to a proposed distinction between two general approaches to symmetries and dualities, called the 'interpretational approach' and the 'motivational approach' (a distinction proposed by Möller-Nielsen 2017). The broad idea is that 'interpretationalists' infer confidently from the mere existence of a symmetry or duality to agreement, i.e. physical equivalence; but 'motivationalists' are cautious: they say that only once a common ontology is formulated should one infer agreement.

Thus Read and Möller-Nielsen are cautious motivationalists. Their position is obviously a cousin of mine—and, I dare say, of De Haro's: as is clear from their helpful comparison of various positions (their section 5.3). But I will not go into detail. For: (i) their definition of duality is different (roughly speaking: weaker) than De Haro's and mine (in our Schema); and (ii) their 'interpretational vs. motivational' distinction concerns the *heuristic* functions of duality for guessing a better theory 'behind' the two duals—functions which I will mostly set aside. But I will agree that these functions are very important (cf. section 3.3.3, *Beware* in section 3.4.1, footnote 7; and De Haro forthcoming). Indeed, these functions will suggest a diagnosis of *why* the false consensus above gelled: i.e. why my Remark has been missed. In short: the diagnosis is that the authors are using the duality to guess a better theory 'behind' the duals, rather than (like me) interpreting dual theories as now formulated.

3.2 A Schema for Duality

In this section I review our account of duality; and in the next, I relate it to ideas of interpretation. This account has been developed mainly by De Haro (2016, section 1; 2020, section 1), but also more fully by us together (2017, sections 2, 3) and foreshadowed in De Haro et al. (2017, section 3). As a mnemonic, we label this account a *Schema*. It will be clear that our Schema is logically weak, so that there are countless examples, including elementary ones: a topic taken up in section 3.3.3.

De Haro and I say that a duality is an isomorphism of theories. More precisely, duality is a matter of:

(a) the two given theories sharing a *common core*; the common core is itself a theory, which we call a *bare theory* (it is usually best to think of the bare theory as uninterpreted, or abstract: though it may be interpreted); and
(b) the two given theories being *isomorphic models* of this common core: that is, isomorphic as regards the structure of the bare theory.

The idea is as in representation theory (in the mathematical sense of 'representation'—a word with confusingly diverse uses, in mathematics and in philosophy): for example, matrix representations of an abstract group G. Each model represents the bare theory, i.e. is a homomorphic copy of it; and although two representations are in general *not* isomorphic, in a duality, the two models in fact are isomorphic.

The two dual, i.e. isomorphic, theories will usually be called *model triples*, the 'triple' referring to the fact that the theory consists of three items: a state-space S, a set of quantities Q, and a dynamics D.[5]

So here, a model is a specific realization—one might say 'version' or 'formulation'—of a theory. That is: it 'models' (verb!) the bare theory. So: *Beware*: the word 'model' often has three connotations that are for our Schema *misleading*. That is: the word 'model', as contrasted with 'theory', often connotes:

(i) a specific solution (at a single time: or for all times, i.e. a possible history) for the physical system concerned, whereas the 'theory' encompasses all solutions—and in many cases, for a whole class of systems;
(ii) an approximation, in particular an approximate solution, whereas the 'theory' deals with exact solutions;
(iii) an interpretation, and even being part of the physical world (in particular, being empirical, and/or observable); whereas the 'theory' stands in need of interpretation, and is of course not part of the physical world.

So NB: *our use of 'model' denies all three connotations.*

[5] Two clarifications, one formal and one interpretative. (1): Agreed: not all theories are best, or usually, presented in this way. In statistical and quantum physics, theories are often formulated in terms of partition functions and/or path integrals. And in spacetime theories, the dynamics is usually presented as field equations holding at each point, not in the 'Schrödinger' and 'Heisenberg' pictures used below. But we believe that our account of duality can be carried over to such theories. (2): Agreed: a theory is almost never presented to us as a tidy triple; so the triple needs to be extracted from a vaguer and/or more complicated presentation. We also make no claim that there is always a best definition of the triple. For more about (1) and (2), cf. De Haro and Butterfield (2018, section 2.1.2).

A standard piece of physics jargon can also be misleading: namely, calling a duality's correspondence between items—the mapping of a state or quantity in one dual theory to the other—a *dictionary*. Philosophers, beware: the 'dictionary' need not be faithful to meanings. Items that it pairs can have very different meanings. Indeed: this will be crucial to cases of (Contr) and (Diff), and thus to the Remark and Implication.

A rigorous but advanced example of this Schema is: *bosonization*, a duality between two quantum field theories in two spacetime dimensions: one with bosons, and one with fermions. So here: a bosonic model triple is isomorphic (by a unitary equivalence) to a fermionic model triple, their common core being a certain infinite-dimensional algebra (De Haro and Butterfield 2017, sections 4, 5). Given how different bosons and fermions are, this is a remarkable, indeed very surprising, isomorphism. But, as it happens, it is *not* a case of (Contr) or (Diff): the two model triples are about a common subject-matter, about which they say different but consistent things.

To give more details, it is clearest to have a notation in which the model triple, comprising the state-space S, the set of quantities Q, and the dynamics D: (i) is separated from the model's own specific structure, and (ii) expresses only the model's realizing (typically: representing, i.e. being a homomorphic copy of) the bare theory T. Thus we sometimes write a model M of T as a quadruple

$$M = \langle S_M, Q_M, D_M, \bar{M} \rangle =: \langle m, \bar{M} \rangle, \tag{1}$$

where: $m := \langle S_M, Q_M, D_M \rangle$ is the model triple; and $\bar{M}$ is the *specific structure*, the 'details', that a model adds to its bare theory.

But note that m is usually built by using the specific structure $\bar{M}$. So m should not be thought of as given independently of $\bar{M}$; and one should not think of M as just the ordered pair of independently given m and $\bar{M}$. Rather, m is built from the specific structure $\bar{M}$: think of building a homomorphic copy of an abstract group G by a specific choice of 7×7 complex matrices. And then m encodes M's representing the bare theory T.

It is clear that—just as in representation theory—two model triples are in general *not* isomorphic to each other, or to the bare theory: in the relevant sense of 'isomorphic', viz. given by the bare theory. So the assertion of duality is substantive: it asserts that two model triples are in fact isomorphic.

Before formally defining duality as isomorphism, we sketch how states, quantities, and dynamics will be treated. Suppose we are given a set of states S, a set of quantities Q, and a dynamics D: $\langle S, Q, D \rangle$. We will write $\langle Q, s \rangle$ for the value of quantity Q in state s. It is these values that are to be suitably preserved by the duality, i.e. by the isomorphism of model triples: cf. eq. 4 below.

For classical physics, one naturally takes quantities as real-valued functions on states, so that $\langle Q, s\rangle := Q(s) \in \mathbb{R}$ is the system's possessed or intrinsic value, in state s, of the quantity Q. For quantum physics, one naturally takes quantities as linear operators on a Hilbert space of states, so that $\langle Q, s\rangle := \langle s|\hat{Q}|s\rangle \in \mathbb{R}$ is the system's Born-rule expectation value of the quantity. (But in fact, for quantum dualities, the duality often preserves off-diagonal matrix elements $\langle s_1|\hat{Q}|s_2\rangle \in \mathbb{C}$: cf. below.)

As to dynamics, here assumed deterministic: this can be presented in two ways, for which we adopt the quantum terminology, viz. the 'Schrödinger' and 'Heisenberg' pictures (though the ideas occur equally in classical physics). We will adopt for simplicity the Schrödinger picture. So we say: D_S is an action of the real line $\mathbb{R}$ representing time on S: $D_S : \mathbb{R} \times S \ni (t, s) \mapsto D_S(t, s) =: s(t) \in S$. There is an equivalent Heisenberg picture of dynamics with D_H, an action of $\mathbb{R}$ representing time on Q: $D_H : \mathbb{R} \times Q \ni (t, Q) \mapsto D_H(t, Q) =: Q(t) \in Q$. The pictures are related by: $\langle Q, s(t)\rangle = \langle Q(t), s\rangle$.

We can now present our Schema for duality as an isomorphism between model triples. Let M_1, M_2 be two models, with model triples $m_1 = \langle S_{M_1}, Q_{M_1}, D_{M_1}\rangle$ and $m_2 = \langle S_{M_2}, Q_{M_2}, D_{M_2}\rangle$. We can suppose that M_1, M_2 are both models of a bare theory T.

To say that the model triples m_1, m_2 are isomorphic is to say, in short, that: there are isomorphisms between their respective state-spaces and sets of quantities, that

(i) make values match, and
(ii) are equivariant for the two triples' dynamics (in the Schrödinger and Heisenberg pictures, respectively).

Thus these maps are our rendering of physics jargon's dictionary.

Thus we say: a *duality* between $m_1 = \langle S_{M_1}, Q_{M_1}, D_{M_1}\rangle$ and $m_2 = \langle S_{M_2}, Q_{M_2}, D_{M_2}\rangle$ requires:

(1) an isomorphism between the state-spaces (almost always Hilbert spaces, or for classical theories, manifolds):

$$d_s : S_{M_1} \to S_{M_2} \quad \text{using } d \text{ for 'duality'; and} \tag{2}$$

(2) an isomorphism between the sets (almost always algebras) of quantities:

$$d_q : Q_{M_1} \to Q_{M_2} \quad \text{using } d \text{ for 'duality';} \tag{3}$$

Fig. 3.1 Equivariance of duality for dynamics: for states and quantities

such that:

(i) the values of quantities match:

$$\langle Q_1, s_1 \rangle_1 = \langle d_q(Q_1), d_s(s_1) \rangle_2 \,, \quad \forall Q_1 \in \mathcal{Q}_{M_1}, s_1 \in \mathcal{S}_{M_1} \tag{4}$$

(ii) and d_s is equivariant for the two triples' dynamics, $D_{S:1}, D_{S:2}$, in the Schrödinger picture; and d_q is equivariant for the two triples' dynamics, $D_{H:1}, D_{H:2}$, in the Heisenberg picture: see Figure 3.1.

Eq. 6 appears to favour m_1 over m_2; but in fact does not, thanks to the maps d being bijections.

De Haro and I have related this Schema to various topics such as: (a) the idea of symmetry, (b) the construction of theories by abstraction from models, and (c) the interpretation of theories; (cf. 2017, sections 3.1, 2.4, 2.3, 3.2.3; 2019; De Haro forthcoming). For this chapter, I only need to develop (c) the interpretation of theories (section 3.3). This will make precise the idea of a subject-matter; and so lead in to presenting examples of the Remark (section 3.4).

3.3 Interpreting Physical Theories

'Interpreting physical theories' is a large subject, at the intersection of logic, especially semantics, and philosophy of science. In this section, I undertake three tasks. I report my endorsement of *intensional semantics* (section 3.3.1). Then I report how that framework makes precise the notion of *subject-matter*—and, thereby, section 3.1's two 'ways to disagree', (Contr) and (Diff) (section 3.3.2). The 'shape' of my Remark will then be clear (section 3.3.3).

3.3.1 Intensional Semantics

This is the framework for analysing the meanings of words and sentences, developed by such authors as Carnap, Montague, and Lewis, that combines Frege's distinction between sense and reference with the tradition, from Leibniz to modern modal logic, of a set of possible worlds, i.e. all the ways the universe could be, only one of which is actual.

At its simplest: a singular term such as 'the Earth's tallest building' has an intension (a Fregean sense) that is a partial function on the set W of worlds mapping a world $w \in W$ to the term's extension (reference) in w—in the example, to the tallest building on Earth, in w. And if w has no Earth, or Earth in w has no buildings, then the function is undefined. Similarly, a predicate such as 'is made of steel' has an intension that is a function sending each world w to the predicate's extension, i.e. set of instances, in w. The semantics of sentences, both simple and compound, is then built up compositionally.

Although this framework was originally developed for formal languages in logic and for (fragments of!) natural languages, it is readily adapted to physical theories with their talk of systems, states and quantities, as in section 3.2. Roughly speaking: systems are objects, i.e. references of singular terms; quantities are relations of objects to numbers (relative to a system of units); a state is a value-assignment to all the quantities that apply to a given object; and a possible world is the total histories of the states of all the objects in the world.

De Haro and I endorse this framework, especially in the version by Lewis (1970). And our endorsement is not just in general, but as applied to section 3.2's conception of theories, models, and dualities. Thus the assignment of intensions and extensions to states and quantities in model triples can be made precise in terms of *interpretation maps*. These are partial functions on the sets of states and quantities, S and Q, into sets of intensions (Fregean senses) or extensions (worldly referents). (Details are given in 2017, section 2.3; and as applied to dual model triples, section 3.2.3.) But this chapter does not need further details about this framework, except in two regards.

(1): The framework has the great merit of respecting the meanings of words! It takes there to be a fact about what is the correct intension of 'the Earth's tallest building': relative of course to various factors, such as choice of language, resolutions of vagueness, context of utterance, etc. That may seem to you a feature so obviously mandatory for an endeavour calling itself 'semantics' that you may ask why I have stressed it. But as we will see in section 3.5, there are 'semantics' that investigate the mathematical consequences of assigning arbitrary meanings (specifically, extensions) to words—viz. the semantics in books of logic, and especially model theory!

Two qualifications about this trenchant remark 'against logic'. (i): It will become clear in section 3.5 that this comment is meant as a cheeky appetizer for my argument, rather than as a criticism of those books. Of course, logic and model theory teach us a great deal about semantics, broadly understood, by studying the consequences of assigning arbitrary meanings. (ii): I agree that it is a good question, perhaps the central question of philosophy of language, exactly what facts make true the correct assignment of intensions. About this, I admire Lewis' answer (1974; 1975; 1983, 370–377). I also admire Callender and Cohen's claim that we should analyse representation in science in terms drawn from philosophy of language and mind (2006, especially section 3). I also maintain that these authors' claims are compatible with section 3.1.1's two morals.

(2): Later on, it will be useful to deploy the framework's precise notion of a subject-matter, as a partition of the set *W* of worlds: to which I now turn.

3.3.2 Subject-Matters: (Contr) and (Diff)

As so often, one cannot say it better than Lewis. He writes:

> We can say that two worlds are exactly alike with respect to a given subject matter. For instance two worlds are alike with respect to the 17th Century iff their 17th Centuries are exact intrinsic duplicates (or if neither one has a 17th Century).
>
> This being exactly alike is an equivalence relation.... The equivalence relation on worlds partitions the worlds into equivalence classes. The equivalence classes are propositions, ways things might possibly be. An equivalence class is a maximally specific way things might be with respect to the subject matter. So a... way to think of a subject matter... is as the partition of equivalence classes.
>
> (Lewis 1988a, 111f)

Thus the idea is that a subject-matter is a taxonomy or classification-scheme, so that worlds that match exactly as regards the taxonomy are to be in the same cell of the partition of *W* (the same equivalence class). Lewis goes on to discuss various ensuing ideas and applications (1988a, 112–117; cf. also 1988b, 136–141). I need only note three of the ideas, and can set aside Lewis' applications, interesting though they are. These three ideas will enable me to state (Contr) and (Diff) more precisely.

First: one subject-matter can *contain*, i.e. be more detailed than, another: like the seventeenth century contains the 1680s, or the subject-matter, how many stars there are, contains the subject-matter, whether there are finitely or infinitely many stars. In terms of partitions, this is a matter of the first partition being a refinement, or fine-graining, of the second. That is: each cell of the second partition is a union of cells of the first.

Second: a proposition is *about*, i.e. entirely about, a subject-matter if its truth-value is determined by the facts about the subject-matter: that is, if the set of worlds at which it is true is a union of cells of the subject-matter.

Third: in general, of course, partitions *cut across* one another, in that neither is a refinement of the other. That is: at least one cell of one partition overlaps, i.e. intersects, two cells of the other. The extreme case, which Lewis calls the subject-matters being *orthogonal*, is: each cell of each partition intersects each cell of the other. An example might be: how many stars there are, as a subject-matter, is orthogonal to Queen Victoria's preferences about clothes. That is: any number of stars is compatible with any sartorial tastes she could have had.

It is clear that even if you are suspicious of possible worlds, and/or of intentional semantics, these ideas will be useful whenever you are concerned with various partitions of some given set. For a philosopher of physics, the prototype is of course partitions of the phase space Γ, in classical mechanics. Each quantity Q, taken as a real function on Γ, defines a partition of Γ whose cells are its level surfaces (inverse images of Q's values). So Q defines a subject-matter. Taking a function, $f(Q)$, of Q is, in general, a coarse-graining. For in general, some two arguments of a function f yield the same value: so that at least one cell of $f(Q)$'s partition is a union of cells from Q's partition. This prototype also gives examples of orthogonality in Lewis' sense. For many classical mechanical systems, any value of position for some given component of the system is compatible with any value of its momentum, so that the component's position and momentum are orthogonal as subject-matters. Similarly, for the same quantity—position, say—on two different components: each component can be anywhere, wherever the other is. Similarly neat examples of subject-matters, and of these related ideas, will be evident in section 3.4's elementary examples of dualities.

I can now use the above summary of intensional semantics and subject-matters to give a more precise statement of section 3.1's two ways that two theories can disagree, (Contr) and (Diff). These formulations will dominate the rest of the chapter.

In section 3.1, I said that for (Contr), two dual theories make contrary assertions about a common subject-matter, so that though they may agree on some claims, they contradict one another over other claims. Given section 3.2's Schema, and the above account of possible worlds, propositions, and subject-matters, we think of conjoining the assertions made by an interpreted model triple into a single proposition, which is made true by a set of worlds. Since the interpreted model triple is about an intended subject-matter (such as—to look ahead to section 3.4.1—point particles, interacting by Newtonian gravity), this set of worlds is a union of cells of the partition that *is* the subject-matter. Thus we have:

- (Contr): Each of two dual model triples is interpreted as wholly true (its conjunctive proposition is wholly true) at a union of cells of a common subject-matter. But these two unions are disjoint: for the propositions contradict each other.

Similarly: in section 3.1, I said that for (Diff), two dual theories describe different subject-matters, though the descriptions are 'isomorphic' or 'matching'—hence the duality. Applying again the notions introduced above, this becomes: the two interpreted model triples are about two different subject-matters (such as—to look ahead to section 3.4.2—position as against momentum). And these subject-matters, i.e. partitions, cut across one another: maybe even in the extreme sense Lewis dubbed 'orthogonality'. Thus we have:

- (Diff): Two dual model triples are interpreted as wholly true (the conjunctive proposition of each is wholly true) at distinct sets of worlds. Each set is a union of cells of the triple's subject-matter, i.e. partition. But the partitions are different, and so are the sets. The sets need not be disjoint: both the model triples could be, both of them, wholly true. But the sets are distinct.

3.3.3 Credo

The 'shape' of my Remark is now clear. By invoking the intended meanings or interpretations of physical theories, I will exhibit cases of (Contr) and (Diff). The isomorphisms given by the duality (eqs. 2 to 4) will not be 'allowed' to rule out such cases by their 'carrying' meanings across to one dual from the other, i.e. their fixing the first dual's interpretation by leaning on the isomorphism so as to use an existing interpretation of the second dual. For carrying meanings across in that way would violate the first dual's intended meaning.

Of course, I do not mean to suggest that two duals' intended meanings are usually, let alone always, different: not all pairs of duals disagree, in my umbrella sense. Indeed: since section 3.2's definition of duality is formal, regardless of meanings, some pairs agree 'utterly'—agree in a stronger sense than do the duals in position-momentum duality, or in bosonization. Thus consider a formulation of relativity theory with a $(+,-,-,-)$ signature for the metric, and another formulation with a $(-,+,+,+)$ signature. (Thanks to Thomas Barrett for pressing this example.) *Prima facie*, the first formulation says that a particle's worldline has positive length, the second that it has negative length. But, of course, everyone concurs that these formulations in no way disagree. The $+/-$ in the metric's signature is a paradigm case of a 'trivial convention'; and so the two formulations, construed with their intended meanings, are indeed equivalent.

Should we then try to strengthen the definition of duality, so that such cases of 'trivial equivalence' are ruled out as not being dualities? I resist this project, because of the first moral in section 3.1.1. The project would require distinguishing 'fact' from 'convention' once and for all, across a whole range of theories (or if you prefer: formulations of theories). And since at least sixty years ago, that has seemed like a mug's game. So: better, say I, to keep to a formal definition of duality, independent of interpretations; and to admit that when classifying an example of it as the duals agreeing ('equivalent'), or as disagreeing (Contr or Diff), we must exercise judgement about what best respects each dual's intended meaning.

Similar remarks apply in connection with the fact that our Schema is not just formal, but logically weak. So it fits countless cases which would not normally be called 'dualities'. And countless such cases would not even be articulated at all—we need no special word for isomorphic representations of any algebraic structure!

But De Haro and I think (2017, sections 2.1, 3.2.2) that it is best to keep to a formal and logically weak definition of 'duality', while admitting that, of course, the *scientific interest* of a duality depends on interpretative issues: in particular, the isomorphism having such features as the following:

(i) being non-obvious, and/or
(ii) being of a logically strong structure (e.g. '10 dimensional Lie group', not just 'group'), and/or
(iii) having other scientific advantages: one important case being that the duality renders the weak coupling, and therefore tractable, regime of one theory dual to the strong coupling, and therefore intractable, regime of the other—so that one can solve 'easy' problems in the tractable regime of the first theory so as to solve 'hard', i.e. intractable, problems in the second theory.[6]

Agreed: one might want to strengthen the definition of 'duality' by requiring such features. As we shall see, feature (i) is often a matter of the notions that correspond by the duality—say, the quantity Q on the 'left', $Q \in Q_{M_1}$, and its image $d_q(Q)$ on the 'right'—being *disparate*. (Recall section 3.2's comment that a duality 'dictionary' need not be faithful to meanings.) Thus: in one of section 3.4's examples, Q is position, while $d_q(Q)$ is momentum.

But it is impossible to be precise about amounts of disparateness, or non-obviousness, or of logical strength. So again: I resist this project of strengthening the definition, because of the morals in section 3.1.1. In any case, this chapter's project does not need any such strengthening: though, as we shall see, the examples that exemplify the Remark have (or, as a matter of history, once had) most of the features (i) to (iii).

[6] These three desiderata are of course widely endorsed, though often with different labels: like 'Striking' and 'Useful' in De Haro et al. (2017, section 2). Thus the famous duality, bosonization, on which our (2017) concentrated has all of them. Here, we return to dualities' *heuristic* importance for guessing new theories, which I mentioned at the end of section 3.1.2, and will take up again in section 3.4.1.

So much by way of a *credo* about what is worth making make precise (even formal), and what is best left to judgement in each case. I now turn to elementary examples of the Schema.

3.4 Examples in Classical and Quantum Physics

In this section, I present two *Examples* of the Schema (though only the second is usually called a 'duality'). The Remark will be evident: two dual theories can disagree, in one of the two ways, (Contr) and (Diff). They can contradict one another about the same subject-matter, or they can be about different subject-matters. Example (1) will illustrate (Contr) (section 3.4.1), and Example (2) will illustrate (Diff) (section 3.4.2).

I will only sketch these Examples. There will be two main limitations of scope. First: it will be clear that there are choices to be made about what exactly is the bare theory, and thus the notion of representation in play in defining the model triples and their specific structure. But we will not need to make very precise choices in order to illustrate the Schema and the Remark. Second, to save space, I will downplay the role of dynamics, i.e. eq. 3.1.

3.4.1 (1): Newtonian Mechanics with Different Standards of Rest

Consider two formulations of Newtonian point-particle mechanics (say, N particles with gravitation) that differ in what they identify as absolute space (what inertial timelike congruence is 'truly at rest').

Here the natural choice for the bare theory would be a neo-Newtonian (also called: 'Galilean') formulation of point-particle mechanics, which postulates a flat affine connection on spacetime, but no absolute rest. Thus the spacetime might be taken to be $\mathbb{R}^4$. Then in the bare theory, the obvious congruence of inertial timelike lines, i.e. the lines $\langle t, x, y, z \rangle$ (with t varying, and x, y, z fixed, for each line), is on a par with—no more 'at rest' than—any congruence boosted with respect to it, such as $\langle t, x - vt, y, z \rangle$ (with t varying, and x, y, z fixed, for each line; and v fixed for the whole congruence, representing a boost in the x direction).

Then the two duals (two models, in our sense) would be formulations of point-particle mechanics set in Newtonian spacetime. So the specific structure by which the two duals differ lies in their specifications of absolute rest. The 'left' dual might specify as absolute rest, the first-mentioned congruence, with lines $\langle t, x, y, z \rangle$, while on the other hand, the 'right' dual specifies $\langle t, x - vt, y, z \rangle$.

We can fill this out a little in terms of the state-space, the map d_s, etc. The configuration space of the N particles can be taken to be $\mathbb{R}^{3N}$. (I say 'can be taken' because to be rigorous: (i) one should instead take the affine space, and (ii) one should remove the coincidence points where two or more particle locations coincide.) Then a state can be an assignment of values of (absolute!) position and velocity to each particle: $(x_1, \ldots z_N; \dot{x}_1, \ldots . \dot{z}_N)$. We may then define the duality map d_s on states so that it takes a state $(x_1, \ldots z_N; \dot{x}_1, \ldots . \dot{z}_N)$ of the first ('left') model to the state in the second ('right') model with the same numerical values, with respect to *its* (the right model's) specification of absolute rest, as the argument-state had for the the left model's specification. (The chapter's motto applies!) Think of Mr Right's absolute rest as boosted with respect to Ms Left's by, say, velocity v in the x direction. This definition of d_s, together with a definition of d_q that sends 'absolute rest' to 'absolute rest' etc., will preserve the values of absolute quantities, i.e. would satisfy eq. 4.

This Example is a case of (Contr), for the duals disagree about what is absolute rest. The values match, i.e. eq. 4 is satisfied, even though d_q pairs Ms Left's and Mr Right's intended meanings for 'absolute rest'—and so respects their mutual disagreement, so that the reference, or extension, of 'absolute rest' is different on the two sides. Namely: the values match, eq. 4 is satisfied, because d_s maps a state of Ms Left to another state that gets the same numerical values for Mr Right's corresponding (by d_q) quantities.

To sum up: the 'left' isomorphic set of states is all the possible instantaneous states of N particles that inhabit a Newtonian spacetime, equipped with 'this' absolute rest; and the 'right' isomorphic set of states is all the possible instantaneous states of N particles that inhabit a Newtonian spacetime, equipped with 'that' absolute rest, defined as boosted with respect to the left, i.e. 'this' one. (I spoke in terms of instantaneous states; but given the two-way determinism, I could have said 'all the possible histories'.) And this pair of duals, on the left and right, illustrates (Contr). In short: we have rendered the empirical undetectability of Newton's absolute space in spacetime terms, as two formulations disagreeing about *which* timelike congruence is at rest—and thus exemplifying (Contr).

Beware: it is tempting to say that there is no real disagreement, that the contrary specifications of absolute rest are 'gauge', or 'a distinction without a difference', or 'a sign that we should move to a neo-Newtonian formulation of point-particle mechanics', a formulation in which the 'surplus structure of absolute rest (i.e. specific structure in our Schema's jargon) is eliminated'.

I agree that it is *tempting* to say these things. But the point is: these temptations are the benefit of hindsight, i.e. of our now knowing the neo-Newtonian formulation. Back in 1715, at the time of the Leibniz-Clarke correspondence, Newton's and Clarke's views were perfectly tenable: recall the remarks in this section's preamble, about a duality being non-obvious at a certain historical epoch. Thus, interpreting the duality with Newton's and Clarke's views, or returning to that earlier epoch: the duality illustrates (Contr).

Agreed: these 'temptations', as I have called them, hint at two other important functions of duality. Namely, to prompt us to guess either

(i) the bare theory, the 'common core', when we have no formulation of it (or only a defective, or intractable, or unperspicuous formulation), or
(ii) another theory 'behind the duals', of which the two duals are not representations (*à la* the Schema) but approximations, in two different parts of some parameter space (different regimes or sectors).

Typically, a theory (ii) will be 'more different' from the duals than is the bare theory (i), because it aims to describe more phenomena than they do: e.g. by including phenomena at higher energies, while each of the duals describes a low-energy regime. (Or replace energy with another important scale, such as velocity or curvature.) Thus for Example (1), with the bare theory (i) taken as a neo-Newtonian formulation, a putative theory (ii) could be general relativity. Clearly, guessing such theories, when one only knows the two duals, is hard!

Obviously, these two heuristic functions of a duality are scientifically very important. But for this chapter, I set them aside, as does much of the philosophical literature about dualities: for it concentrates on dualities' connections with issues in logic and ontology, such as theoretical equivalence (as this chapter does) and symmetry, rather than issues in methodology.[7]

Apart from illustrating (Contr), this Example also has four noteworthy Features. We will see these Features vary, or recur, in the next subsection's Example: for Features (1) and (2) will bear on whether an Example illustrates (Contr) or (Diff):

(1) *Toy cosmology?* It is natural to think of each dual (and the bare theory) as a 'toy cosmology': not least because of the historical influence of point-particle world-pictures, as advocated by such figures as Boscovitch and Laplace. That is: it is natural—or at least, it was traditional—to take each formulation as saying that (a) there are no bodies apart from the N particles (i.e. there is no environment of the system—which is the cosmos!) and that (b) there are no forces apart from the given Galilean-invariant one, such as Newtonian gravitation.

[7] Exceptions include: Dawid (2013; 2017, sections 4.2, 4.3), Rickles (2011, 66; 2013, 319; 2013a, 67–78; 2017, 66), De Haro's discussion of (i) and (ii) for gauge/gravity duality (2016, section 2.1; 2016a, section 2.1; 2017, section 3.2), and his general discussion of the heuristic function (2018). The latter distinguishes (i) and (ii): De Haro calls (i) a 'theoretical function' (especially section 4.1.2), and calls only (ii) a 'heuristic function', and its theory a 'successor theory' (sections 4.1.3 and 5).

(2) *Complete set of states?* In each dual, we have what the tradition, the textbook, would regard as the complete set of states for the 'toy cosmos' that are compatible with the specification of absolute rest that has been made.
(3) *The size of the duality group?* In this Example, there is a whole boost group worth of duals, i.e. $\mathbb{R}^3$-worth. For any vector in $\mathbb{R}^3$ can represent the velocity with which to boost from a fiducial specification of the absolute rest. But we focus on just two duals. (In the jargon of *duality groups*: we take the duality group to be Z_2.)
(4) *More variety*: Of course, Newtonian point-particle mechanics, as usually understood, encompasses varying N, while we have implicitly taken a fixed value of N. But this was just to keep matters simple. That is: the simplest strategy is to have each bare theory (and so its models) fix N, and to express the intended generality of the original theory by simply universally quantifying over N. But maybe one could formulate a more abstract bare theory that encompasses all N.

Note also that one could write down several Examples that are 'cousins' of Example (1). Thus one could write down theories other than point-particle mechanics, e.g. fluid mechanics, with two contrary specifications of absolute rest. The natural choice for the bare theory would again be a formulation of e.g. fluid mechanics using a neo-Newtonian spacetime. And the two duals would be formulations set in a Newtonian spacetime that disagree with each other about absolute rest. These duals will again illustrate (Contr). And broadly speaking, the discussion in Example (1) about the state space, the definition of the duality maps, and the four noteworthy Features will carry over, *mutatis mutandis*.[8]

3.4.2 (2): Position-Momentum Duality in Elementary Quantum Mechanics

I turn to the unitary equivalence, by Fourier transformation, between the position (i.e. Schrödinger) representation and the momentum representation in elementary wave mechanics. For simplicity, we consider a single spinless non-relativistic particle in one spatial dimension. Thus one might take the bare theory to be the Hilbert space $L^2(\mathbb{R})$, equipped with its algebra of (say, bounded) linear operators.

[8] I say 'broadly speaking' because there will be some substantial differences. For a start, the configuration space of a fluid is infinite-dimensional. So the technical details of the states, quantities, and duality maps will be much more intricate. Also, Feature (1) falters, in that we do not normally think of a sample of a classical fluid as the entire material content of the universe: but agreed, this may just reflect the historical contingency that no such world-picture was as influential as its point-particle 'rivals' by such figures as Boscovitch and Laplace.

Then the idea is: the left dual is fixed by the choice of position, the right by the choice of momentum.

It is usual to think of these choices as just 'choices of basis'. More precisely, so as to respect the fact that position and momentum are continuous quantities: as 'choices of a spectral family of projectors'. The unitary Fourier transformation is then a passive transformation sending an arbitrary physical state, expressed in position representation, to the momentum representation of the same physical state: and so preserving all quantities' expectation values. A sharply peaked wave function $\psi : \mathbb{R} \to \mathbb{C}$ is mapped to a spread-out one, and vice versa, attesting to the uncertainty principle. But we interpret it as the identity function on physical states, since we define a corresponding map on physical quantities (represented as linear operators) in such a way that any quantity's expectation value is preserved. (For details, cf.: at a pedagogic level, e.g. Jordan 1969, section 18; or more rigorously, Prugovecki 1981, sections 4.3, 4.5; Takhtajan 2008, 89–92.)

In terms of our notations, d_s is the identity map on physical states, and d_q is the identity map on physical quantities. So of course eq. 4, thus interpreted, is satisfied. But neither map is the identity map on states', or quantities', expressions as, respectively, complex functions on $\mathbb{R}$, and linear operators on such functions. Interpreting eq. 4 in terms of these expressions, it expresses the equality of an inner product calculated in unitarily related orthobases.

But this usual way of thinking of the duality does *not* serve my present purpose, viz. exemplifying the Remark. For it clearly does not make the left and right duals disagree, in either of our senses (Contr) or (Diff). The two duals' claims *do* agree, though they are expressed in different languages ('bases').[9]

However, we *can* get an illustration of (Diff) by thinking of this duality differently. We need 'to think like Bohr': that is, to take the position and momentum 'perspectives' ('windows', 'contexts') to exclude each other, where 'exclude' means not so much 'contradict' as 'cannot be adopted together'.

Thus the left dual is to be *only* about position. We can take its set of states to be just the set of probability distributions for position; so in an elementary wave-mechanical formalism, a state is a probability density given as the squared modulus of the wave-function: $p(x) \equiv |\psi(x)|^2$. And the left dual's quantities are to be just position Q (i.e. multiplication by the argument-variable x) and the (Borel) functions of position, $f(Q)$: their expectation values are then, in the elementary formalism, $\langle f(Q) \rangle = \int dx\ f(x)p(x)$. Similarly, the right dual is only about momentum. Its states are just the probability distributions for momentum,

[9] But this usual way of thinking is an important prototype for other dualities where the claims of the left and right duals *agree*. *Bosonization* is an advanced example, again with a unitary equivalence (De Haro and Butterfield 2017, sections 4, 5).

$p(k)$, where we write the argument variable as k to signal the interpretation as momentum; and its quantities are just the (Borel) functions of momentum, $f(P)$, with expectation values $\langle f(P) \rangle = \int dk\, f(k)p(k)$.

We now define the duality map d_s to map the state $p(x)$ to the *same* mathematical real function: but now interpreted as a probability density for momentum (so that we again write the argument variable as k). So the mathematical duality map d_s is the identity map on probability distributions on $\mathbb{R}$: but as interpreted on physical states, it is an active transformation—*not* the identity map. Similarly for quantities: each Borel function f is mapped to itself, as a pure mathematical function, but the interpretation changes: $d_q : f(Q) \mapsto f(P)$.

With these definitions, the 'preservation of values' in our Schema, eq. 4, is *not* a preservation of physical, interpreted expectation values, as in the usual way of thinking of the duality (again, the motto is illustrated). Rather, d_s and d_q preserve the 'shape' or 'graph' of mathematical real functions p (a probability density) and f (a Borel function).

It is clear that this gives a case of (Diff). For the duals describe different subject-matters: position and momentum, respectively. It is also clear that, unlike Example (1), there is *no* temptation—even in hindsight, when one has long been accustomed to the duality— to say that there is no real difference between the duals, or that the contrast between position and momentum is 'gauge', or 'a distinction without a difference', or 'a sign that we should move to a mechanics that eliminates the contrast'.

Let us compare this with Example (1), as regards the four noteworthy Features. For the first three, we get the opposite verdict to that in Example (1); for the fourth, the Examples are similar:

(1) *Toy cosmology?* It is not natural to think of each dual (or the bare theory) as a toy cosmology. There might well be an environment to the system. In particular, there could be another replica system, another spinless non-relativistic quantum particle confined to another (non-intersecting) spatial line. And we could study our original given system via its position probability distributions, and the replica system via its momentum distributions. If so, the two duals would 'describe different subject-matters' in two senses of that phrase: both different quantities and different systems.

(2) *Complete set of states?* In each dual, we *do not* have what the tradition, the textbook, would regard as the complete set of states for the system. There is only position information on the left, and only the momentum information on the right. On each side the 'other information', contained in the phase of the wave-function, has been erased by taking the modulus squared.

(3) *The size of the duality group?* Given the left dual, defined as, say, the spectral family for position, and the idea of a unitary transformation from it to a right dual, there is a vastly richer set of definable right duals than just the $\mathbb{R}^3$-worth we saw in Example (1). After all, the unitary transformations on $L^2(\mathbb{R})$ are an infinite-dimensional set.

(4) *More variety*: Again, the idea of the duality encompasses a generality I have not yet mentioned. The quantum particle that is the system can be in two or three spatial dimensions, instead of one; and there can be 1, or 2, or... N particles. Again, the simplest strategy is to keep the bare theory reasonably concrete (hence our choice of $L^2(\mathbb{R})$), and to express the intended generality by simply universally quantifying, i.e. having many close-cousin dualities.

To sum up this section: These two examples establish my main Remark: duals can disagree with each other, either by:

- (Contr): making contrary assertions about a common subject-matter, such as the standard of absolute rest, as in Example (1); or
- (Diff): making claims about distinct but suitably 'isomorphic' subject-matters, such as the continuous quantities, position and momentum, in quantum mechanics (Example 2).

So, as announced in section 3.1.1, my task now is to spell out the Implication of this Remark for proposed definitions, in logic and philosophy, of theories being 'equivalent' (section 3.5).

3.5 An Implication about Theoretical Equivalence

The Remark has a simple implication—I shall say 'Implication'—for accounts of theoretical equivalence. Namely: two dual theories satisfying (Contr) or (Diff) might get formalized so as to be logically equivalent—where I use 'formalized' and 'logically equivalent' with their standard meanings, in logic and philosophy. But obviously such duals are not equivalent—as 'equivalent' is normally understood. So such cases show that logical equivalence is too weak an explication of theoretical equivalence. And so also, therefore, is any weakening of logical equivalence.

As I explained at the end of section 3.1.1, this Implication, though simple, is worth stressing. For the recent philosophical literature on theoretical equivalence has focused on various weakenings of logical equivalence: a focus motivated by logical equivalence being clearly, in one regard, too strong. So it is worth seeing that—notwithstanding the literature's rationale for weakening logical equivalence—there is also a problem 'in the other direction'. That is: a defect of being too weak, rather than too strong (as an explication of theoretical equivalence).

I will fill out this Implication in three subsections. The first is a warning about jargon, especially 'theoretical equivalence' and 'physical equivalence' (section 3.5.1, which returns us to section 3.3.1's advertisement for intensional semantics). The second spells out the Implication for logical equivalence (section 3.5.2). The third shows that the Implication also holds for notions weaker than logical equivalence, which have been proposed in the literature (section 3.5.3).

3.5.1 A Warning about Jargon

We saw at the end of section 3.1.1 that 'theoretical equivalence' is a term of art. While all agree that we need some notion of 'making the same claims about the world', there is no agreed definition. (This is like 'proposition' in the philosophy of logic and language: all agree we need some notion of 'meaning of a sentence', but the exact definition varies or is lacking.) We also saw that logical equivalence, as a candidate definition, is both:

(i) too strong, because of cases of synonymy (like 'electric' and 'électrique'), prompting the effort to give a weaker definition; and
(ii) too weak, because of cases of homonymy (like 'bank' in finance and beside a river): i.e. cases where logical equivalents disagree in their claims about the world—which will be illustrated by my Remark's Implication.

Agreed, you might want to reserve the phrase 'theoretical equivalence' for some weakening of the familiar logical equivalence. If so, you might classify the pairs of theories the Remark draws attention to as theoretically equivalent. (It will depend on your exact definition of 'theoretical equivalence'.) But, of course, the implication of the Remark remains. Namely, expressed without the jargon of 'theoretical equivalence', as interpreted, two such dual theories do not make the very same claims about the world.

And to reflect the Remark, it will of course be natural (to avoid confusion) to introduce another phrase for 'making the very same claims about the world'. As mentioned in section 3.1.2, De Haro (2016, section 1.2; 2016a, section 1.2) suggests *physical equivalence*: a phrase adopted by us (2017, sections 1.1, 3.2.3(1)), and also by others, e.g. Huggett (2017, especially 85f).

Fair enough, but *beware*: this phrase can also be misleading. For in philosophy, we often distinguish the vast plethora of all (logically and/or metaphysically) possible worlds, W, from its subset, the set N of all the nomically possible worlds: the worlds that obey the actual laws of nature. (At least, we right-thinking Humeans, believing that the laws of nature are contingent, do so.) And the nomically possible worlds are often called the 'physically possible' worlds, in a nod to the success of physics in formulating our best guesses about the laws of nature. Thus, with *this* jargon in mind, one naturally hears 'physical equivalence' as meaning 'equivalence across all the physically possible worlds', i.e. across N: which, since $N \subset W$, will be a *weaker* relation than 'equivalence across all the possible worlds', i.e. across W. This means that the label 'physical equivalence', for our notion of 'making the very same claims about the world', can be misleading, since my point—the Implication—is that some pairs of theories, when interpreted, make different claims and yet are logically equivalent.

And finally: another confusing feature of the various jargons is that the logic books' 'logical equivalence' is not really the same as 'equivalence, i.e. sameness of truth-value, across all the worlds': even when we take W as the set of all logically possible worlds, so that we naturally hear 'logical equivalence' as equivalence across all of W. For again: the logic books' 'logical equivalence' is formal—tied to a vocabulary, and disregarding any intended meanings—and so will rule e.g. 'The weather is fine' and 'Il fait beau' as inequivalent.

This last point returns us, of course, to section 3.3.1's advertisement for intensional semantics. *That* is a framework for analysing meaning, etc. that respects intended meanings: and according to which 'same meaning', i.e. our desired notion of 'making the same claim about the world', indeed corresponds to 'equivalence across all the worlds'. More power to its elbow...

3.5.2 The Implication: For Logical Equivalence

The Remark—dual theories can disagree with each other—implies a limitation of standard approaches for formalizing physical theories (indeed, as we shall see, theories about any topic). For two such disagreeing dual theories might get formalized so as to be equivalent 'within the meaning of the act', i.e. equivalent according to the approach's precise definition of 'equivalence'. Or to put it more charitably, since 'theoretical equivalence' is, after all, a term of art (section 3.5.1): the Remark implies that these approaches' use of 'theoretical equivalence' has a misleading connotation.

This Implication follows because the duality implies that for this mishap, this misleading formal equivalence, to occur, the only thing needed is appropriate homonymies; that is, the same word being used in the two theories, but with mutually 'inverted' meanings—meanings that of course formalization sets aside.

Recall section 3.2's warning about the physics jargon 'dictionary': so now the dictionary will pair the same word with mutually inverted meanings, rather than different words with the same meaning, as 'dictionary' usually connotes. (But not all the words used in the theories need to have their meanings 'inverted' by the dictionary: just an appropriately chosen subset.)

I shall first explain this Implication for the most well-known approach to defining 'theoretical equivalence': namely, logic books' notion of logical equivalence. The next subsection explains how the Implication applies to other approaches.

Thus the logic books take theories as sets of sentences (closed under logical deduction); and then define theories as *logically equivalent* if they are stated in the same language (i.e. have the same non-logical vocabulary: also known as 'signature'), and are made true by the same interpretations, i.e. have the same set of models (in the sense of 'model' used by logic and semantics, rather than *our* sense in section 3.2!). (Cf. e.g. Hodges 1997, 37f.) So for this approach, the Implication is that two dual theories that disagree with one another (in sense Contr or Diff) may well get formalized so as to be logically equivalent, because their different claims about the world are 'isomorphic claims'.

We can make vivid this Implication, this mishap, in terms of the ubiquitous Alice on the left, i.e. advocating the left dual, and Bob on the right, advocating the right dual. If Alice and Bob are 'unwise enough' to use the same language in their advocacy of their doctrines—in the *syntactic* sense of 'language'—and an appropriate subset of the words they both use have appropriately 'inverted' meanings, then what they *say*, in the syntactic sense, i.e. the words they utter, could be the same, despite their theories disagreeing. In this case their theories might well get formalized, i.e. regimented into a formal language, so as to render the resulting formal theories logically equivalent.

Thus take Example (2), position-momentum duality (Diff: section 3.4.2). Suppose Alice advocates the left dual: she is a 'position monomaniac'. But she speaks standard English, so that her word 'position' means position. And suppose Bob advocates the right dual: he is a 'momentum monomaniac'. But the semantics (intended interpretation) of Bob's language is 'position-momentum' inverted with respect to Alice's, i.e. with respect to standard English. So Bob uses the word 'position' to mean momentum. Then, thanks to the duality, Alice and Bob in advocating their different theories may say the very same set of sentences. So the theories might get formalized so as to be rendered logically equivalent. Similarly, of course, for Example (1), absolute rest (Contr: section 3.4.1).

We can also state the Implication in terms of the duality maps d_s etc.: and thereby illustrate our motto about how confusing language can be! Thus in Example (1): suppose Alice describes a state s by saying 'the Nth particle is at rest'. Then Bob describes the transformed state $d_s(s)$ (i.e. boosted relative to Alice by the same amount as is his specified standard of rest) by saying the same words. But they contradict one another: (Contr).

In Example (2), Alice describes a state s by saying 'the particle has a Gaussian distribution for position centred at $x = 5$'. Bob describes the transformed state $d_s(s)$ (i.e. the same probability distribution but for momentum instead) by saying the same words. Their assertions differ, but are compatible: (Diff). For 'position' in Bob's mouth means momentum (in Alice's mouth, and ours). And Bob could be referring to a different particle.

So much by way of explaining the Implication, as applied to the notion of logical equivalence. There are two comments to make.

(A) *Permutation arguments*: Mention of homonymy, and especially of appropriately 'inverted' meanings, is bound to remind philosophers of other permutation arguments, such as the inverted spectrum in the philosophy of perception (e.g. Shoemaker 1982; Putnam's argument for the indeterminacy of reference, Putnam 1981, chapter 2, appendix; Lewis 1984; Button and Walsh 2018, chapter 2), and Einstein's hole argument (connected to Putnam's argument by Rynasiewicz 1994). A rich field for comparison with this chapter's project—but one for which I have no space.

(B) *The labour of formalization*: My saying 'might well get formalized' is shorthand for some substantial labour. For by the standards of logic, physical theories are almost always presented very informally. Even formulations that by the standards of theoretical physics are rigorous are a far cry from what a logician would call a formal theory. Even the multi-variable calculus of Newtonian point-particle mechanics (Example (1)), let alone the functional analysis needed for rigorous quantum theory (Example (2)), are difficult to capture adequately and perspicuously in formal languages, especially with a first-order logic. (At the end of section 3.5.3, we will see that this trouble is one motivation for recent proposals about theoretical equivalence.)

Nor is it just a matter of difficulty, or of physicists' and logicians' different habits or interests as regards rigour. There are matters of principle. For as we learn in logic, continuity is a second-order notion. That is (roughly speaking): to express the difference between the reals and the rationals—whether every bounded set has a least upper bound—you need second-order quantification: which brings with it the incompleteness of second-order logic. (The usual response is to stay in first-order languages, and adopt, following Tarski, a first-order axiom scheme for continuity: e.g. Goldblatt 2004, 61f.; Andréka et al. 2012, 643 and references cited there.)

So my saying 'might well get formalized' is programmatic. But notice that the labour and obstacles just summarized have nothing to do with dualities. So it remains very plausible that two duals could get formalized so as to be logically equivalent: thanks to the isomorphism of model triples and the easy correspondence between the 'knobs' (e.g. specifications of absolute rest) that yield (Contr) and (Diff).

3.5.3 The Implication: For Weaker Notions of Equivalence

Logical equivalence is of course just one notion of equivalence between theories. This prompts the question: how does the Implication fare with other notions? In this section, I will argue that since the other proposed notions of 'theoretical equivalence' are weakenings of logical equivalence, the Implication again holds good.

I begin by recalling that the logical tradition of taking a theory as a set of sentences, which leads to defining logical equivalence as above, has been rejected by many philosophers of science. They advocate instead a 'semantic conception' of theories, saying that a theory is a set of models: though authors differ about what they mean by 'model', and what are the merits of the semantic conception. Obviously, space prevents a full discussion of this conception, and even of how the Implication fares with it. I will make just four comments about the Remark and Implication. It is the fourth that leads to proposed weakenings of logical equivalence, and so is the most important of the four.

(1) *Better off? Prima facie*, the semantic conception seems able to accommodate the Remark's cases of (Contr) and (Diff). For it promises to classify two such duals as being 'different theories', simply because section 3.2 had different model triples on the two sides. So this suggests that this conception faces no 'embarrassing Implication' from the Remark's cases of (Contr) and (Diff).

(2) *Not so fast!* On the other hand, our Schema allows dualities whose duals do *not* disagree. Witness, in quantum theory, position-momentum duality and bosonization (cf. the start of section 3.4.2 and footnote 9). And more generally: the semantic conception will individuate theories far too finely if it declares two theories different whenever there are different models (or classes of models). For whatever exactly one means by 'model', a model will have 'specific structure' (cf. section 3.2) that may not contribute to its representational role. So overall, the semantic conception will have some work to do to account for dualities—at least according to our Schema.[10]

[10] Halvorson (2012, section 4.2, 193–197) argues in general, without regard to dualities, that the semantic conception will individuate theories far too finely if they are identified with a set of models. But van Fraassen (2014, 278–281) replies that the semantic conception never endorsed so fine an individuation of theories. Cf. footnote 3.

(3) *A false contrast?* The point in (2), about the danger of individuating theories far too finely, recalls more general difficulties facing the semantic conception, some of which erode the distinction between it and the syntactic conception of theories which was its target. In brief, this goes as follows. (Cf. Halvorson and Tsementzis 2017 and Lutz 2017 for recent detailed critiques of the contrast between syntactic and semantic conceptions.) In order for the semantic conception to describe models' representational role, and free itself from 'specific structure', it is drawn into associating words with the representational parts of models. But for model theory (in logic's sense!), and thus also for the syntactic conception, a model is tied to a specific non-logical vocabulary (predicates, etc.) that gets assigned referents in the model's domain. As van Fraassen (1989, 366) put it: "models are defined, as in many standard logic texts, to be partially linguistic entities, each yoked to a particular syntax". So given any class of such models all tied to the same vocabulary, there is a well-defined set of sentences true in each model, which model theory calls the 'theory' of the class. Agreed, this theory need not be axiomatizable, or cast in a first-order language: but the syntactic conception was never committed to theories having either of those features.

(4) *Weakening logical equivalence.* As I have announced: the literature on theoretical equivalence (whether adopting the syntactic conception or the semantic one) has focused on a limitation of defining 'theoretical equivalence' as logical equivalence that goes in the *opposite direction* than does our Remark. That is: there are pairs of theories that:

   (a) we intuitively think of as the same, but that:
   (b) are not logically equivalent, in the above sense.

   Informal mathematics gives many simple examples that are more precise than my mention of electromagnetism in English and in French. Examples: (i) the theory of Boolean algebras vs. the theory of Boolean rings; (ii) topology as axioms governing a notion of 'open set', vs. as axioms governing 'closed set', vs. as axioms governing 'closure operation'. Such pairs of theories show logical equivalence is too strong as a criterion, or explication, of 'theoretical equivalence'.

Of course, this limitation has long been recognized by logicians and philosophers, including advocates of the syntactic conception. And so began the hunt for a logically weaker explication. For this hunt, Weatherall (2019) is an excellent introduction; early and recent references in logic are de Bouvere (1965), Visser (2006), and Andréka and Németi (2014); and recent references in philosophy are Barrett and Halvorson (2016a,b, 2017), Button and Walsh (2018, chapter 5), Hudetz (2016, 2019a,b), and Weatherall (2016, 2018).

But this is not the place to review the ideas and results of this literature. I will just sketch one main idea, called *definitional equivalence* (also called 'synonymy', as in my own jargon), and then emphasize that it makes no difference to the Remark and Implication. That is: dual theories that satisfy (Contr) or (Diff), as in the Remark, might well get formalized so as to be definitionally equivalent. It will then be clear that two other proposed weakenings of logical equivalence also make no difference to the Remark and Implication.

The idea of definitional equivalence is that starting from one presentation of a theory—say, Boolean algebras in terms of 'meet' and 'join' in a lattice—one can introduce definitions of the other presentation's notions (e.g. operations yielding a ring) and then rigorously deduce all the claims, i.e. theorems, of the other presentation; and vice versa, starting with the other presentation. Such deductions are almost always conducted in the informal mathematical natural language of a textbook. But there is a strong consensus in mathematics about which proofs are rigorous enough that they could be made wholly rigorous, e.g. by being transcribed into a predicate calculus formulation of a standard set theory such as Zermelo–Fraenkel set theory (ZFC).

Accordingly, definitional equivalence is made precise by mathematical logic (with its deliberately setting aside knowledge of intended meanings!) by taking the two presentations as formal theories, i.e. as sets of sentences in a formal language, closed under deduction: call them T_1 and T_2. To avoid trouble from ambiguous vocabulary items, e.g. the same predicate occurring in both languages, one assumes that the vocabularies of T_1 and T_2 are disjoint. Then T_1 and T_2 are *definitionally equivalent* iff: (i) one can add to T_1 a definition of each vocabulary item of T_2, in such a way that within this augmentation of T_1 one can deduce all of T_2; and of course (ii) vice versa. Thus (i) is called: making a *definitional extension* of T_1, and showing T_2 to be (a sub-theory of) a definitional extension of T_1. So: definitional equivalence is a matter of the two theories each being definitionally extendable so as to 'encompass' the other. This is also called: their having a *common definitional extension*.

It is clear, without revisiting the details of our Examples, that this notion of definitional equivalence makes no difference to the Remark and Implication. And for the now-familiar reason, viz. the logical framework deliberately sets aside intended meanings. Thus suppose that the advocates, Alice and Bob, of such a pair of dual theories are 'wise enough' to use disjoint vocabularies. Nevertheless, their theories might well, once formalized, be definitionally equivalent. Indeed, for a duality in the sense of our Schema, their theories no doubt *will* be definitionally equivalent, if they could have been rendered logically equivalent, by the argument of section 3.5.2, had Alice and Bob used the same vocabulary. Thus to repeat the Remark: despite the definitional equivalence, Alice and Bob are still making different claims about the world!

Essentially the same point comes up in the literature on theoretical equivalence, regardless of dualities. It is folklore among logicians that often, formal theories that they (anyone!) would think of as very different are definitionally equivalent: in effect, because the definitions added to one theory of the predicates, etc. of the other are allowed to be 'whatever it takes' to deduce the theorems of the other theory. That is: the definitions are not required to be 'intuitive' or to 'respect meanings'. Hudetz (2016, example 2; 2019a, proposition 4) gives an example which is striking, especially in the light of our previous discussion. He points out that a formal theory of Minkowski spacetime (postulating the affine space of $\mathbb{R}^4$, with a Minkowski interval) is definitionally equivalent to a formal theory of Euclidean geometry for $\mathbb{R}^4$. The reason is that they are each a definitional extension of the theory of the real line $\mathbb{R}$. Roughly speaking: one 'just' uses set theory to form appropriate n-tuples and to define an inner product with the desired signature. Then each can be definitionally extended to 'encompass' the other. From within Minkowski geometry, one defines a second inner product with completely positive signature $(+, +, +, +)$; and vice versa, from within Euclidean geometry.

This is striking in two ways. (i): It illustrates the general idea that any two definitional extensions of a given theory that introduce different terms are definitionally equivalent since they can each add the other's definitions to yield a common definitional extension. (ii): Recall section 3.3.3's example of two formulations of relativity theory with trivially swapped conventions about $+/-$ in the metric's signature (used to emphasize that dual theories might well agree utterly in their claims). Here, on the other hand, we naturally think of definite vs. indefinite signature as encoding important geometric differences: differences that the formal notion of definitional equivalence is blind to. Thus, taking the two examples together, we have a striking illustration of how interpretation moulds verdicts of theoretical equivalence.

Finally, I note that my Remark and Implication still hold good, beyond the case of definitional equivalence. To be specific: here are two other weakenings of logical equivalence that have been proposed, and that have evident merits as explications of theoretical equivalence:

(A) *Generalized definitional equivalence*. This generalizes, i.e. weakens, the idea of definitional equivalence to deal with many-sorted logic (Barrett and Halvorson 2016a,b, who call it 'Morita equivalence'). This generalization seems right, in that it enables us to say, for example, that a formulation of Euclidean geometry based on points, i.e. with points in the domains of its models, is equivalent to ('the same theory as') a formulation based on lines (Barrett and Halvorson 2017).

(B) *Categorical equivalence*, i.e. equivalence of categories in the sense of category theory. Note that this notion is defined for *any* two categories. So the broad idea here is to argue that a scientific theory is best conceived as a category: the models of the theory (in some sense of 'model'!) are the objects of the category, and the arrows of the category are appropriate maps between models. This is not the place for details about this broad idea, for which cf. Halvorson and Tsementzis (2017), Weatherall (2017). I just remark that on the syntactic conception of theory, one takes the objects to be models in the sense of model theory, i.e. as in section 3.5.2, and the arrows to be elementary embeddings. On this approach, (A) implies (B) but not conversely (cf. Barrett and Halvorson 2016a, section 5 and 6, especially theorems 5.6, 5.7).

What matters for this chapter is whether the Remark's Implication still holds. As discussed: it is bound to hold for a notion of theoretical equivalence that is weaker than logical equivalence—which definitional equivalence is; and so is (A); and so is (B), when (B) is implemented on the syntactic conception of theory, as above. And the Implication still holds for the now-familiar reason: the definitions of theoretical equivalence, by (A) and by (B) on the syntactic conception, are formal—they set aside intended meanings.

To sum up this section: the Implication is worth emphasizing because:

(1) the drift of the literature on 'theoretical equivalence' is to weaken the notion from the elementary conception of logical equivalence, while
(2) the Implication of the Remark is that in cases of dual theories, *logical equivalence is too weak*. Its verdict 'these theories are equivalent' misses the fact that the theories, as interpreted, disagree with each other.

3.6 Envoi

I have argued, on the basis of De Haro's and my proposed account of duality as isomorphism, that two theories can be dual, though we interpret them as disagreeing—either by contradicting each other or by making claims about different subject-matters (albeit consistent claims). So the theories, thus interpreted, are not equivalent. I went on to discuss an implication of this verdict: that as a criterion of theoretical equivalence, logical equivalence is too weak—as is, therefore, any weakening of it.

Furthermore, I pointed out that these conclusions bear on 'philosophy beyond spacetime', in two ways: ways that correspond to whether our philosophical interest in a duality centres on (i) the interpretation of the duality, as at present stated, or (ii) its heuristic function—what the duality suggests for 'tomorrow's theory'.

Thus as to the first interest, (i): I pointed out that string theory has (admittedly conjectural) dualities, whose dual theories can be interpreted as disagreeing (in either of the above ways). Indeed, they disagree about basic features of space, such as its dimension or the radius of a compact dimension (cf. footnote 1). Thus insofar as current string theory gives us a glimpse of the correct theory of quantum gravity, the ramifications of these conclusions for our understanding of spacetime are considerable. To spell this out: on the interpretations of the dual theories that I claim are tenable, we would need to decide which of the two duals to commit to. In one example, this would be a matter of committing to one dimension of space, rather than another; in the other example, it would be a matter of committing to one value rather than another, for the radius of a compact dimension of space.

As to the second interest, (ii), I admitted (e.g. at the end of section 3.1.2) that this chapter has downplayed the heuristic function of dualities. But I suggested that recent authors' views that two string-theoretic dual theories should be interpreted as agreeing, *contra* my claims, may be explained, at least in part, by these authors' predominant interest being in dualities' heuristic function. (Cf. my warning *Beware* in section 3.4.1 and its footnote 7.)

Agreed: these philosophical interests, (i) and (ii), of course interact. Think of how 'committing to one interpretation' under (i) raises the usual issues about scientific method, and in particular about scientific realism concerning today's (or at least tomorrow's) string theory. These are issues that are central in assessing the heuristic function of dualities, i.e. to (ii).

And such connections prompt an obvious final comment about the prospects for future work in 'philosophy beyond spacetime'. Namely: despite this chapter having not engaged with speculations about quantum gravity, in particular dualities in string theory, I hope that carefully addressing interpretative matters about simpler dualities can serve as a useful preliminary to such speculations.[11]

[11] This chapter is dedicated to Graeme Segal: what an inspiration he is, to heart as much as to mind. I am very grateful to Sebastian De Haro. This chapter owes a lot to his ideas (and patient explanations!). For comments and correspondence, I am grateful to Graeme Segal; to the vertices of the Cambridge Simplex, here anonymized *à la* Bourbaki; to E. Curiel and N. Dewar for detailed help (I only wish I could have acted on it all!); and to T. Barrett, R. Dawid, N. Huggett, L. Hudetz, O. Pooley, J. Read, B. Roberts, and C. Wüthrich. I also thank audiences in Cambridge, Dubrovnik, New York, Oxford, and Utrecht; and the editors, not least for their patience.

References

Andréka, H., J. X. Madarász, I. Németi, and G. Székely (2012). A logic road from special relativity to general relativity. *Synthese 186*(3), 633–649.

Andréka, H. and I. Németi (2014). Definability theory course notes. Department of Logic, ELTE, Budapest, https://old.renyi.hu/pub/algebraic-logic/DefThNotes0828.pdf.

Barrett, T. and H. Halvorson (2016a). Glymour and Quine on theoretical equivalence. *Journal of Philosophical Logic 45*(5), 467–483.

Barrett, T. and H. Halvorson (2016b). Morita equivalence. *Review of Symbolic Logic 9*(3), 556–582.

Barrett, T. and H. Halvorson (2017). From geometry to conceptual relativity. *Erkenntnis 82*, 1043–1063.

Button, T. and S. Walsh (2018). *Philosophy and Model Theory*. Oxford: Oxford University Press.

Callender, C. and J. Cohen (2006). There is no special problem about scientific representation. *Theoria 21*, 67–85.

Coffey, K. (2014). Theoretical equivalence as interpretative equivalence. *British Journal for the Philosophy of Science 65*, 821–844.

Dawid, R. (2013). *String Theory and the Scientific Method.* Cambridge: Cambridge University Press.

Dawid, R. (2017). String dualities and empirical equivalence. *Studies in History and Philosophy of Modern Physics 59*, 21–29.

de Bouvere, K. (1965). Synonymous theories. In J. W. Addison, L. Henkin, and A. Tarski (Eds.), *The Theory of Models: Proceedings of the 1963 International Symposium at Berkeley*, pp. 402–406. Amsterdam: North Holland Publishing Company.

De Haro, S. (2016). Duality and physical equivalence. http://philsci-archive.pitt.edu/id/eprint/12279. (This is an expansion of De Haro 2020; and the title has changed to 'Spacetime and Physical Equivalence'.)

De Haro, S. (2018). *Theoretical equivalence and duality.* In preparation.

De Haro, S. (forthcoming). Theoretical equivalence and duality. *Synthese*. https://doi.org/10.1007/s11229-019-02394-4.

De Haro, S. (2020). Spacetime and physical equivalence. In N. Huggett, K. Matsubara, and C. Wüthrich (Eds.), *Beyond Spacetime: The Foundations of Quantum Gravity*, pp. 257–283. Cambridge: Cambridge University Press.

De Haro, S. and J. Butterfield (2017). A schema for duality, illustrated by bosonization. In J. Kouneiher (Ed.), *Foundations of Mathematics and Physics One Century After Hilbert*, pp. 305–376. New York: Springer.

De Haro, S. and J. Butterfield (2019). On symmetry and duality. *Synthese*. https://doi.org/10.1007/s11229-019-02258-x.

De Haro, S., N. Teh, and J. Butterfield (2017). Comparing dualities and gauge symmetries. *Studies in History and Philosophy of Modern Physics 59*, 68–80.

Goldblatt, R. (2004). *Orthogonality and Spacetime Geometry*. New York: Springer.

Halvorson, H. (2012). What scientific theories could not be. *Philosophy of Science 79*, 183–206.

Halvorson, H. and D. Tsementzis (2017). Categories of scientific theories. In E. Landry (Ed.), *Categories for the Working Philosopher*, pp. 402–429. Oxford: Oxford University Press.

Hodges, W. (1997). *A Shorter Model Theory*. Cambridge: Cambridge University Press.

Hudetz, L. (2016). Definable categorical equivalence: Towards an adequate criterion of theoretical intertanslatability. Manuscript, October.

Hudetz, L. (2019a). Definable categorical equivalence. *Philosophy of Science 86*(1), 47–75.

Hudetz, L. (2019b). The semantic view of theories and higher-order languages. *Synthese 196*(3), 1131–1149.

Huggett, N. (2017). Target space ≠ space. *Studies in History and Philosophy of Modern Physics 59*, 81–88.

Jordan, T. (1969). *Linear Operators for Quantum Mechanics*. Mineola, NY: Dover Publications Inc.

Lewis, D. (1970). General semantics. *Synthese 22*(1/2), 18–67. Reprinted in his *Philosophical Papers: Volume 1* (1983), Oxford: Oxford University Press.

Lewis, D. (1974). Radical interpretation. *Synthese 27*(3), 331–344. Reprinted with Postscripts in his *Philosophical Papers: Volume I* (1983), Oxford: Oxford University Press.

Lewis, D. (1975). Languages and language. In K. Gunderson (Ed.), *Language, Mind, and Knowledge*, pp. 3–35. Minneapolis, MN: University of Minnesota Press.

Lewis, D. (1983). New work for a theory of universals. *Australasian Journal of Philosophy 61*(4), 343–377. Reprinted in his *Papers in Metaphysics and Epistemology* (1999), Cambridge: Cambridge University Press.

Lewis, D. (1984). Putnam's paradox. *Australasian Journal of Philosophy 62*(3), 221–236. Reprinted in his *Papers in Metaphysics and Epistemology* (1999), Cambridge: Cambridge University Press.

Lewis, D. (1988a). Relevant implication. *Theoria 54*, 161–174. Reprinted in his *Papers in Philosophical Logic* (1998), Cambridge: Cambridge University Press; page references to reprint.

Lewis, D. (1988b). Statements partly about observation. *Philosophical Papers 17*, 1–31. Reprinted in his *Papers in Philosophical Logic* (1998), Cambridge: Cambridge University Press; page references to reprint.

Lutz, S. (2017). What was the syntax-semantics debate in the philosophy of science about? *Philosophy and Phenomenological Research 95*, 319–352.

Matsubara, K. (2013). Realism, underdetermination and string theory dualities. *Synthese 190*, 471–489.

Möller-Nielsen, T. (2017). Invariance, interpretation and motivation. *Philosophy of Science 84*, 1253–1264.

Prugovecki, E. (1981). *Quantum Mechanics in Hilbert Space: Second Edition*. Cambridge, MA: Academic Press.

Putnam, H. (1962). The analytic and the synthetic. In H. Feigl and G. Maxwell (Eds.), *Scientific Explanation, Space and Time*, pp. 358–397. Minneapolis, MN: University of Minnesota Press.

Putnam, H. (1981). *Reason, Truth and History*. Cambridge: Cambridge University Press.

Quine, W. V. O. (1951). Two dogmas of empiricism. *Philosophical Review 60*(1), 20–43. Reprinted in his *From a Logical Point of View: Nine Logico-Philosophical Essays* (1980), Cambridge, MA: Harvard University Press.

Read, J. (2016). The interpretation of string-theoretic dualities. *Foundations of Physics 46*, 209–235.

Read, J. and T. Möller-Nielsen (2020). Motivating dualities. *Synthese 197*, 263–291.

Rickles, D. (2011). A philosopher looks at string dualities. *Studies in History and Philosophy of Modern Physics 42*(1), 54–67.

Rickles, D. (2013). AdS/CFT duality and the emergence of spacetime. *Studies in History and Philosophy of Modern Physics 44*(3), 312–320.

Rickles, D. (2017). Dual theories: 'Same but different' or 'different but same'? *Studies in History and Philosophy of Modern Physics 59*, 62–67.

Rynasiewicz, R. (1994). The lessons of the hole argument. *British Journal for the Philosophy of Science 45*, 407–436.

Shoemaker, S. (1982). The inverted spectrum. *Journal of Philosophy 79*(7), 357–381.

Stein, H. (1992). Was Carnap entirely wrong, after all? *Synthese 93*, 275–295.

Takhtajan, L. (2008). *Quantum Mechanics for Mathematicians*. Providence, RI: American Mathematical Society.

van Fraassen, B. (1989). *Laws and Symmetry*. Oxford: Oxford University Press.

van Fraassen, B. (2014). One or two gentle remarks about Hans Halvorson's critique of the semantic view. *Philosophy of Science 81*, 276–283.

Visser, A. (2006). Categories of theories and interpretations. In A. Enayat, I. Kalantari, and M. Moniri (Eds.), *Logic in Tehran: Proceedings of the Workshop and Conference on Logic, Algebra and Arithmetic, Held October 18–22, 2003*. Boca Raton, FL: CRC Press.

Weatherall, J. O. (2016). Are Newtonian gravitation and geometrized newtonian gravitation theoretically equivalent? *Erkenntnis 81*, 1073–1091.

Weatherall, J. O. (2017). Categories and the foundations of classical space-time theories. In E. Landry (Ed.), *Categories for the Working Philosopher*, pp. 329–348. Oxford: Oxford University Press.

Weatherall, J. O. (2019). Part 1: Theoretical equivalence in physics. *Philosophy Compass 14*, e1259.

4
From Quantum Entanglement to Spatiotemporal Distance

Alyssa Ney

Abstract

There is an influential research program in quantum gravity developing the connection between quantum entanglement and spatiotemporal distance. Through a series of well-confirmed results, it has been shown how these facts about the entanglement entropy of component systems may be connected to facts about spatiotemporal distance. Physicists are seeing these results as yielding promising methods for better understanding the emergence of (the dynamical) spacetime from more fundamental quantum theories, and for the development of a nonperturbative theory of quantum gravity. However, to what extent does the case for the entanglement entropy-distance link provide evidence that spacetime structure is nonfundamental and emergent from nongravitational degrees of freedom? I will show that a closer look at the results lends support only to a weaker conclusion: that the facts about quantum entanglement are constrained by facts about spatiotemporal distance, and not that they are the basis from which facts about spatiotemporal distance emerge.

4.1 Introduction

Within the field of quantum gravity, there is an influential research program developing the connection between quantum entanglement and spatiotemporal distance. Quantum information theory gives us highly refined tools for quantifying quantum entanglement. A common measure is the von Neumann entropy S, which can be seen as a quantum analog of the Shannon entropy (Preskill 1998, Nielsen and Chuang 2016). For a state with density matrix ρ,

$$S(\rho) = -\operatorname{tr}(\rho \log \rho). \tag{4.1}$$

Alyssa Ney, *From Quantum Entanglement to Spatiotemporal Distance* In: *Philosophy Beyond Spacetime: Implications from Quantum Gravity*. Edited by: Christian Wüthrich, Baptiste Le Bihan, and Nick Huggett, Oxford University Press.
 DOI: 10.1093/oso/9780198844143.003.0004

Where the density matrix is diagonalizable, with eigenvalues λ_i, the von Neumann entropy may also be expressed as

$$S(\rho) = \sum_i \lambda_i \log \frac{1}{\lambda_i}. \tag{4.2}$$

The von Neumann entropy gives one a measure of how entangled a system is. So, for example, if we consider a bipartite system σ_{AB}, we may ask how entangled are its parts, σ_A and σ_B. In this case, we can refer to $S(\rho_A)$ and $S(\rho_B)$ as the parts' entanglement entropies, where ρ_A and ρ_B are the reduced density matrices of the system's parts. Suppose ρ_{AB} is a pure state. The von Neumann entropy for any pure state is identically zero. If the joint state of the system is a product state so that $\rho_{AB} = \rho_A \otimes \rho_B$, then the entanglement entropy for either part, $S(\rho_A)$ or $S(\rho_B)$, will also be zero. On the other hand, for an entangled system, the entanglement entropy of the parts will not be zero. Indeed, depending on the degrees of freedom of the system, there will be a maximum value of the entanglement entropy associated with the maximally entangled state. For a system of qubits, this will be the value associated with the Bell states, $\psi_{AB} = \frac{1}{\sqrt{2}}(|01\rangle \pm |10\rangle)$ and $\psi_{AB} = \frac{1}{\sqrt{2}}(|00\rangle \pm |11\rangle)$.

Through a series of well-confirmed results, it has been shown how these facts about the entanglement entropy of component systems may be connected to facts about spatiotemporal distance.[1] Physicists are seeing these results as yielding promising methods for better understanding the emergence of (the dynamical) spacetime (of general relativity) from more fundamental quantum theories, and moreover as promising for the development of a nonperturbative theory of quantum gravity. These results appear to be part of a more general convergence forming from several different approaches to quantum gravity (string theory, spin foams, causal sets) around the idea that spacetime, dynamical spacetime of the kind found in general relativity, is an emergent phenomenon, a phenomenon that arises in some sense or other from a more fundamental, nongravitational description (Carlip 2012, Huggett and Wüthrich 2013, Oriti 2021). However, to what extent does the case for the entanglement entropy-distance link provide evidence that spacetime structure is nonfundamental and emergent from nongravitational degrees of freedom? I will show that a closer look at the results lends support only to a weaker conclusion: that the facts about quantum entanglement are

[1] We may note, in anticipation of this fact, that entanglement entropy will be connected with spatiotemporal distance, and that this quantity has certain features that make it apt to serve as a distance relation itself. In particular, it satisfies the triangle inequality: For a joint state ρ_{AB}, $S(\rho_{AB}) \geq |S(\rho_A) - S(\rho_B)|$ (Nielsen and Chuang 2016, 515–516). On the other hand, since, for example, $S(\rho_{AB})$ can equal zero even when $\rho_A \neq \rho_B$, the entanglement entropy has other features that make it seem unlike a distance relation.

constrained by facts about spatiotemporal distance, and not that they are the basis from which facts about spatiotemporal distance emerge.

4.2 Constructing the Metric from Entanglement Entropy in the AdS/CFT Context

Let's start by looking at results connecting facts about the spacetime metric with facts about quantum entanglement that have been motivated from within the string theory paradigm. These follow first on Maldacena's (1999) discovery of the AdS/CFT correspondence. Maldacena shows how one can establish a holographic correspondence between pairs of theories.[2] For, example one may consider first a gravitational theory, a string theory on anti-de Sitter (AdS) spacetime in $d + 1$ dimensions; and next a theory without gravitational degrees of freedom, a conformal field theory, a type of quantum field theory with a kind of scaling symmetry (CFT), defined on a flat spacetime in d dimensions (see Fig. 4.1).

In this model, one may view the AdS description as defining a bulk spacetime for which the CFT lies on its boundary. Maldacena shows how one can translate back and forth between the description of the $(d+1)$-dimensional spacetime in the bulk and the d-dimensional spacetime on its boundary. This work has been extremely influential and so it is a good place to start to look for mainstream arguments on the link between quantum features and spacetime structure. Even though we don't ourselves inhabit an AdS spacetime, this correspondence shows us how one might, for at least one case, derive facts about a curved spacetime, and hence gravity, from a quantum theory without it.

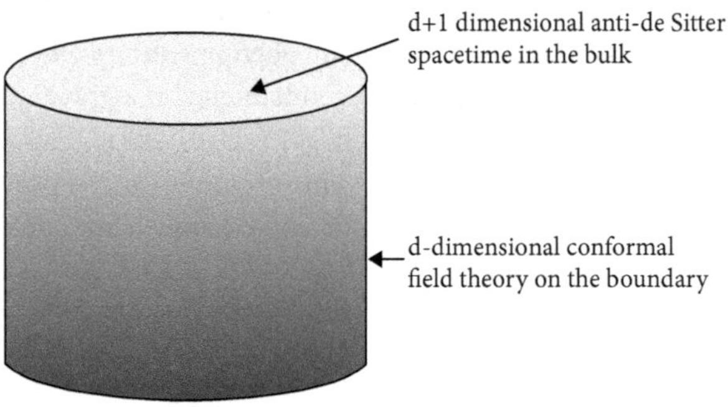

Fig. 4.1 AdS/CFT correspondence

[2] The explicit connection with holography was made in Witten (1998).

I want to look at a particular line of research that has started within the AdS/CFT context, work that aims to derive facts about the AdS metric using particular facts about the CFT on the boundary; in particular, facts about the entanglement entropy of particular states living on the boundary.[3] In 2006, Ryu and Takayanagi asked us to consider the following. First, take a CFT boundary and divide it into two regions, A and its complement, $\bar{A}$ (see Fig. 4.2).

We may then consider the reduced density matrices ρ_A, which describes the state on region A, and $\rho_{\bar{A}}$, which describes the state on the complement region $\bar{A}$. Given these density matrices, we can then compute their entanglement entropy. Assuming the total boundary region is in a pure state, $S(\rho_A) = S(\rho_{\bar{A}})$. Now consider the minimal surface running through the AdS bulk that connects the two edges dividing region A from its complement. Call this surface γ_A (see Fig. 4.3).

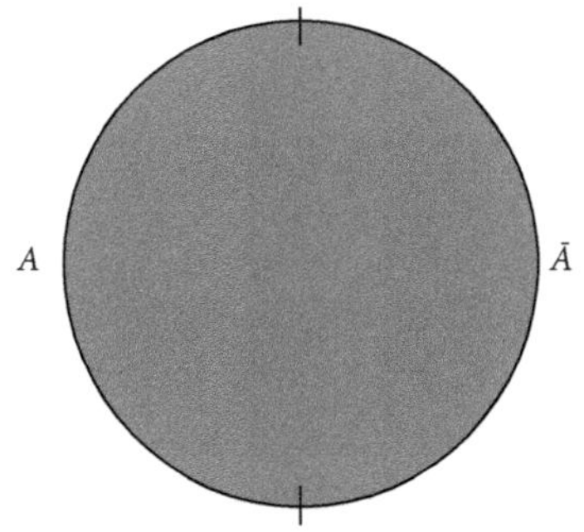

Fig. 4.2 Ryu–Takayanagi cut

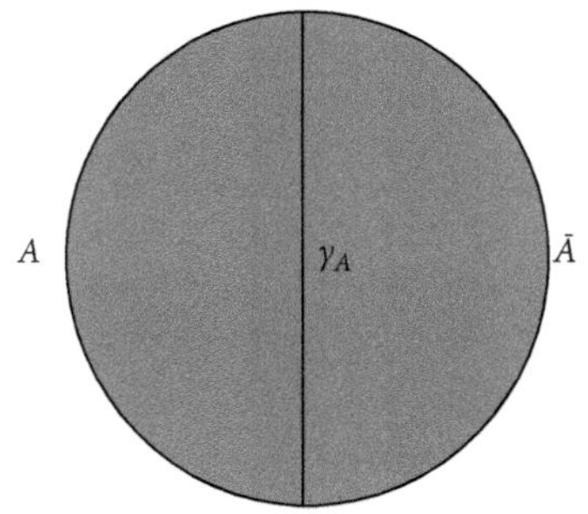

Fig. 4.3 Minimal surface

[3] Rickles (2013), Teh (2013), Dieks et al. (2015), Read (2016), De Haro (2017, 2020), Butterfield (2021), and Huggett and Wüthrich (forthcoming) are prior philosophical works on the AdS/CFT correspondence. Jaksland (2020) discusses the Ryu–Takayanagi conjecture.

The Ryu–Takayanagi conjecture is that a particular relationship exists between the area (or in the $d = 1$ case, length) of γ_A and the entanglement entropy of the region on the boundary A:

$$S(\rho_A) = \frac{\text{area of } \gamma_A}{4G_N^{(d+1)}}, \tag{4.3}$$

where $G_N^{(d+1)}$ is Newton's constant for the relevant $(d + 1)$-dimensional space.[4] Those familiar with work on black hole thermodynamics will immediately recognize that this relationship has the same form as the Bekenstein–Hawking formula that relates the entropy of a black hole to the area of its event horizon (Bekenstein 1973, Hawking 1975):

$$S_{BH} = \frac{\text{area of the event horizon}}{4G}. \tag{4.4}$$

This is something we will come back to. Ryu and Takayanagi motivate their conjecture by appeal to this earlier work on black hole entropy, and in particular the explanation of why this formula holds developed by Sorkin and his collaborators (Sorkin 1983, Bombelli et al. 1986).

We will discuss below how Ryu and Takayanagi establish their conjecture. For now, it is just worth noting how influential this work has been in motivating a research program in quantum gravity, working within the AdS/CFT context, for seeing the connection between facts about quantum entanglement (entanglement entropy) in a quantum field theory without gravitational degrees of freedom, and facts about the metric in a curved (AdS) spacetime. To begin, Hubeny et al. (2007) generalized the result from one about a static region of the AdS spacetime to more general regions, providing a covariant account of AdS distances in quantum information theoretic terms. Later work asked whether it was possible to recover more of the AdS metric from the entanglement properties of regions on the boundary, beyond the area of the minimal surfaces γ_A. That is, how far can we succeed in 'bulk reconstruction', i.e. in gaining an understanding of "how these degrees of freedom, which are well-understood from the perspective of nongravitational quantum field theory, reorganize themselves (in an appropriate limit) into a manifestly local gravitational theory" (Bao et al. 2019, 1)? Indeed, for a complete reconstruction, we would want not only to extract the metric of the bulk, but also the dynamics.

The work on 'hole-ography' of, for example, Balasubramanian et al. (2014), Headrick et al. (2014), and Myers et al. (2014) aims to construct the metrical properties of closed surfaces embedded in the AdS space from entanglement

[4] This is expressed using units where $\hbar = c = k_B = 1$.

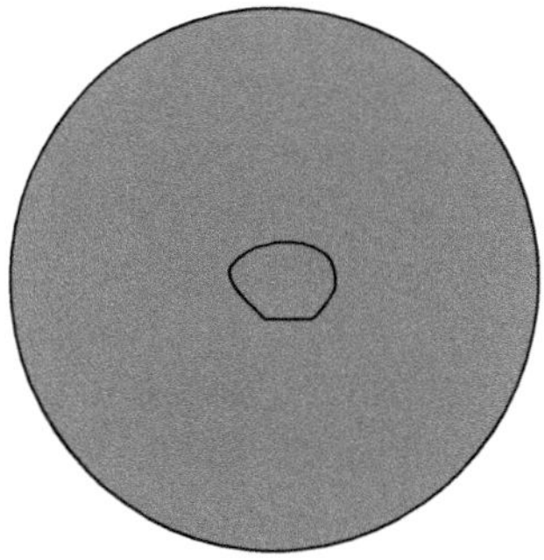

Fig. 4.4 Closed surface in the AdS bulk

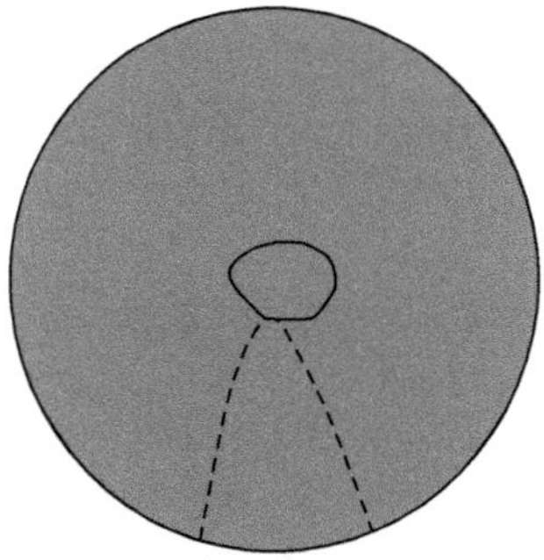

Fig. 4.5 Tangent curve extended to the CFT boundary

entropies on the boundary field theory. The strategy can be (loosely) explicated as follows. Take a closed surface in the AdS bulk (see Fig. 4.4).

Now find geodesics k that are tangent to each region on that closed surface and extend them out to the CFT boundary to get a boundary region I_k (see Fig. 4.5). The idea then is to argue that one can recover the surface's area in terms of the entanglement entropies on the boundary using:

$$E = \sum_{k=1}^{n} [S(I_k) - S(I_k \cap I_{k+1})], \tag{4.5}$$

where E is what Myers et al. (2014) refer to as the differential entropy. The area of the surface is identified with the limit of this quantity as $n \to \infty$.[5]

[5] Differential entropy is one of many entanglement measures that may be used to probe the area of such a bulk surface. Hubeny (2014) argues that it has several limitations. In particular, as Myers et al. (2014) acknowledge, the original proposal is only well defined for three-dimensional, pure AdS geometries, and only there where the bulk surface is not too 'wiggly'. Myers et al. modify the proposal

These developments led Keeler and her collaborators to argue that entanglement entropies on the boundary are sufficient to uniquely fix the metric (up to diffeomorphism) everywhere in the neighborhood of the extremal surfaces (Bao et al. 2019). Although for the generic case, one will not be able to rule out the existence of 'entanglement shadows' in the AdS bulk, regions where the metric is left undetermined by the entropy of boundary regions, perhaps because there are regions which extremal surfaces can't cross because of singularities (cf. Headrick et al. 2014, 38), large steps have already been taken to capture a lot of information about the AdS metric in terms of quantum entanglement on the CFT boundary. Explicit reconstructions have already been successful for bulk spacetimes with large-scale symmetries (Rangamani and Takayanagi 2017).

4.3 The Entanglement Entropy-Distance Link Beyond AdS/CFT: Area Laws

As noted, our universe doesn't have an AdS geometry, and so one might ask whether these results are confined to the AdS/CFT context, or whether we may have justification for seeing a connection between entanglement entropy of the parts of a system and distance in the more general case. Let us back up and reframe a bit what we are after. So far, what we are seeing is the establishment of a series of area laws. Although in classical thermodynamics we are familiar with the idea that entropy is an extensive quantity, scaling with volume, this is not so for the cases we have been considering. In the case of black hole thermodynamics and the Ryu–Takayanagi conjecture, (Bekenstein–Hawking or entanglement) entropies scale with area. And this is what allows us to relate facts about entanglement entropy to facts about the metric.

The work we have just discussed argues for a direct correspondence between the entanglement entropy of a CFT state *A* on the boundary and the area of surfaces through the AdS bulk separating *A* from its complement. We should probe whether this connection between the entanglement entropy of a region *A* and the area of the surface separating it from its complement may be generalized from the AdS/CFT context to more general quantum systems. Indeed, there are indications that area laws may hold more generally outside of the AdS/CFT context, and thus there is a research program devoted to finding the conditions under which they hold. Again, this is inspired by the Bekenstein–Hawking law for black hole entropy.

to apply to more generic spacetimes, but Hubeny stresses that even the revised measures are defined only on static geometries. Her paper (2014) proposes a more robust measure: the covariant residual entropy. This is inspired by the earlier discussion of residual entropy in Balasubramian et al. (2012).

Although it is familiar from the consideration of Bell states in quantum mechanics that quantum entanglement/correlations may persist over indefinitely long spatial distances, interactions in quantum many body systems are typically local. That is, subsystems typically interact over short distances with only a finite number of neighboring subsystems. There is an intuition that interactions are short-range because these correlations are enacted through the interface surfaces[6] between regions. An area law analogous to what we've already seen provides support for this intuition.

In a survey article, Eisert et al. (2010) develop the case for area laws by illustration with systems representable as a lattice, in particular one-dimensional systems where the entire system may be represented as a chain of quantum systems L, and we focus on the entanglement entropy of a part of this chain, e.g. the block I consisting of its first n members: $\{1,\ldots n\}$. If the entanglement entropy scaled with volume, then this would mean that the entanglement entropy of I, $S(\rho_I)$ would be proportional to the cardinality of the block, i.e. n. However, it may be shown instead, for a many body quantum system, that $S(\rho_I)$ scales with the area of the boundary.[7]

For such a one-dimensional chain, the boundary is very simple. The boundary is one or two parts of the block, depending on the boundary conditions imposed. And so, we obtain an area law for the entanglement entropy just in case $S(\rho_I)$ is independent of the block size and instead is of order 1:

$$S(\rho_I) = O(1) \tag{4.6}$$

Eisert et al. emphasize that their claim is not that area laws always hold for many body quantum systems. Rather, one fact determining whether or not an area law holds is whether the system is at a quantum critical point or not. For systems near criticality, the entanglement varies with the length of the chain. And so, what we have is not an area law, but a volume law. This is not surprising given what is known about the breakdown of locality in critical systems in thermodynamics. In what follows, I summarize the argument, eliding most of the technical details.

Consider a bosonic harmonic chain. In this case, the elements of the chain are modeled as discrete versions of complex Klein–Gordon fields; that is, as spinless

[6] I use 'interface surface' here to make a distinction with the 'boundary surface' in the AdS/CFT context. What we discuss here as the interface surface corresponds to the black hole's event horizon or Ryu and Takayanagi's minimal extremal surface.

[7] It is worth noting that the notion of area here is not quite the same as in the case of the Ryu–Takayanagi area law. The area in question here is that of the boundary between the two subsystems. In the latter case, it is the area of the dual bulk surface that ends on the boundary between the two entangled subsystems.

particles with charge. We write down a Hamiltonian for this system in terms of the canonical operators x and p, and system operators X and P which describe the coupling structure. It turns out that even in such one-dimensional systems, simple though they are, the entanglement entropy is quite difficult to compute. However, there is another entanglement measure, the logarithmic negativity E_N, that is more tractable and which provides an upper bound for the entanglement entropy (at least for pure states).

Following Eisert et al., consider the simplest case where the chain L is bisected symmetrically into two parts. We will assume the Hamiltonian describes parts of the system as only able to directly interact with their nearest neighbors, so interactions are local. Here we have a gapped system, one in which there are discrete energy gaps above the ground state that depend on the coupling strength. In this case, Audenaert et al. (2002) proved that:

$$S(\rho_I) \leq E_N(\rho, I) = \frac{1}{2}\log\left(\frac{\|X\|^{1/2}}{\lambda_{\min}^{1/2}(X)}\right), \tag{4.7}$$

where $\|X\|$ is the norm of the system operator X. This was the first rigorous area law for a quantum lattice system. As one can see, the upper bound for entanglement entropy does not depend on the size of the subsystem I. It is thus independent of the block size. This result has been extended to simple higher-dimensional systems, e.g. for quasi-free systems of bosons or fermions arranged on a lattice. For these systems, unlike for the one-dimensional case, the boundary between subsystems is nontrivial. Again, one can calculate the logarithmic negativity using system operators to find that the entanglement entropy scales with the boundary area. Thus, Eisert et al. draw from this work the conclusion that under the assumption of local interactions, entanglement tracks the area of the surface separating the two subsystems. As they put it, describing the import of (4.7):

> This result suggested that the locality of the interaction in the gapped model is inherited by the locality of entanglement, a picture that was also confirmed in more generality. (2010, 5)

In the case of criticality, just as for thermal systems, we lose the locality of interactions, and so the assumptions used for deriving an area law fail. Entanglement entropy, in such cases, scales with volume, as we would otherwise expect by analogy with the thermal entropy.

More recent work by Cao et al. (2017) builds on these proofs of area laws using them (again, outside of the AdS/CFT context) to argue that we can see the spacetime metric as emerging from facts about entanglement entropy. Or, more accurately, Cao et al. (2017) use the quantum analog of the mutual information,

$I(A : B)$, which is defined in terms of the entanglement entropy, and link this with facts about spatial distance (Nielsen and Chuang 2016, 514):

$$I(A : B) \equiv S(\rho_A) + S(\rho_B) - S(\rho_{AB}). \tag{4.8}$$

The spacetime metric for Cao et al. emerges from relations of mutual information between parts of the total quantum state in Hilbert space. Eliding the details, we can just note that again, the inference isn't direct, but mediated by a claim about local interactions:

> The mutual information between regions is a measure of how correlated they are. It provides a useful way of characterizing the 'distance' between such regions *because of its relation to correlation functions between operators*. We expect that in the ground state of a field theory, correlators of field operators will decay as exponentials (for massive fields) or power laws (for massless ones). The mutual information may reflect this behavior, as it provides an upper bound on the correlation function between two operators. (2017, 4, my emphasis)

So again, for Cao et al., it is in the assumption of a local field theory that the relevant area laws obtain. We will return to this locality assumption shortly.

4.4 From Correspondence to Emergence?

The work described so far is very suggestive and, even if this is not always explicit as it is in the work of Cao et al., the establishment of such links between entanglement measures and metrical features has held out the promise for many that one can see spacetime as emerging from facts about entanglement entropy. We certainly have, in these several cases, a bridge that can take us, in certain well-defined contexts, back and forth between entanglement facts and facts about spatiotemporal distance or area. But is there reason to see the one set of facts as emerging from the other?[8]

Rickles (2013), Teh (2013), and Crowther (2014), in earlier discussions of the AdS/CFT correspondence, have warned us to be cautious before moving from

[8] As is standard in this literature, I will continue to use the language of emergence. But for these purposes, I could just as easily have framed the discussion in terms of whether we should see there being a reductive link, whether the facts about spacetime metric are reducible to facts about quantum entanglement. The central question is whether we should view the spatiotemporal facts not merely as derivable from facts about quantum entanglement, but as arising from the facts about quantum entanglement, whether the facts about quantum entanglement provide the explanation of in virtue of what the spatiotemporal facts obtain.

the existence of such a duality or bridge to a claim of emergence. And this is important: although the work we have discussed may enable us to move back and forth between two theoretical frameworks, this is compatible with the holding of several possible inter-theoretic relationships:

1. Facts about the metric emerge from fundamental facts about entanglement entropy.
2. Facts about the entanglement entropy emerge from fundamental facts about the metric.
3. Facts about the metric and entanglement entropy are equally fundamental.
4. Facts about the metric and entanglement entropy are neither fundamental, but both emerge from some more fundamental, as yet undiscovered, framework.[9]

Following Teh, we must recognize that, before we can be justified in adopting a claim of emergence (of facts in theory B from facts in theory A), we must establish that the A-facts are more fundamental—that they provide the explanation of in virtue of what the B-facts arise. Teh (2013) acknowledges that in the AdS/CFT context, often, because it is easier to define the quantum field theory than the gravity theory, there is reason to view the CFT as more fundamental (cf. Horowitz 2005). But on the other hand, others are sympathetic to the possibility that the AdS and CFT theories both emerge from some more fundamental theory (Teh 2013, 309).

Another issue that needs to be considered that is relevant to the question of emergence concerns whether it is reasonable to believe the AdS/CFT correspondence, and in particular the connection between entanglement entropies and spacetime quantities, is exact. Dieks et al. (2015, 208) note that although the AdS/CFT duality is often treated as an exact correspondence, it is not clear whether this is actually so.[10] If it is not exact, then experiments could, in principle, decide the question of which of the two theories is the more fundamental and settle a question of emergence. In the case in which it turns out the correspondence is exact, Dieks et al. are more skeptical that what we have is a situation in which either theory is more fundamental:

> in this case that the two theories completely agree on everything that is physically meaningful, the two sides of the duality should be taken as different representations of one and the same physical theory. (2015, 209)

[9] This is the alternative both Rickles (2013) and Crowther (2014) express sympathy for when discussing the AdS/CFT correspondence.

[10] Harlow (2020) argues that this is confused, and that the AdS/CFT duality is exact. What is not exact, Harlow notes, is the relationship between the facts on either side of this duality and facts in some low-energy effective gravitational field theory.

Today, there is no established consensus on whether one side of the AdS/CFT duality is more fundamental, or indeed whether neither is more fundamental than the other. And so, I argue, we shouldn't be basing a judgment of which of (1)–(4) is correct on what is the scientific consensus about AdS/CFT. Moreover, some have argued that even if it is reasonable to think the AdS/CFT correspondence supports a claim of emergence of spacetime from a more fundamental quantum theory, it doesn't itself explain *how* dynamical spacetime might emerge from a quantum field theory on flat spacetime (cf. Van Raamsdonk 2010). It is precisely here that concepts from quantum information theory are supposed to come to the rescue. The connection with quantum information theory may provide insights on which of (1)–(4) best describes the metaphysical structure of our world, insights that may generalize beyond the AdS/CFT context.

4.5 Gluing with Entanglement?

As just mentioned, Van Raamsdonk (2010) notes that just from the AdS/CFT correspondence alone, we can't see how dynamical spacetime emerges from a nongravitational quantum field theory. However, he believes that if one appeals to quantum information theoretic facts about entanglement measures, then it can be made clear how what we have is a case of emergence. His argument works by considering two cases. The first involves a set of disconnected AdS spacetimes that we can then regard as connected by virtue of the entanglement relations in their dual CFTs. The second case involves a single vacuum CFT in which we gradually remove the entanglement between its component degrees of freedom. As this happens, we can see the corresponding regions in the dual AdS spacetime as becoming pushed away from each other until, when their mutual information goes to zero, the distance between them becomes infinite, and they effectively pinch apart.

These cases are extremely suggestive of the idea that there is a genuine explanatory connection between the metrical features of an AdS spacetime and the entanglement entropies of its dual quantum field theory. And especially in Van Raamsdonk's discussion of the first case, we see a good reason to think that there isn't merely a correlation between the two theories, but that one set of facts is arising because of the other. However, as I will explain, it is not clear to me that these cases are illustrating emergence as in Option (1) of the previous section.

Let's begin with the second case first, as it's simpler, consisting of just one CFT and its dual global AdS spacetime. Here, Van Raamsdonk's discussion of the case provides a fascinating picture of how the entanglement of degrees of freedom in a CFT track distance in its holographic dual (see Fig. 4.6).

Consider a point x_A located on the edge of the bulk near the midpoint of region A on the boundary, and another point $x_{\bar{A}}$ located on the other edge of the bulk

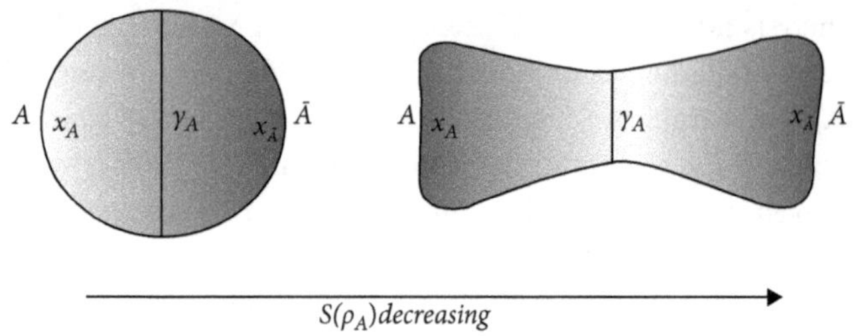

Fig. 4.6 Minimal surface variation

near the midpoint of region $\bar{A}$. What we see is that as the entanglement between region A and its complement is decreased, the area of the region through the bulk AdS spacetime separating A from $\bar{A}$, what we earlier denoted γ_A, decreases, and the distance between points x_A and $x_{\bar{A}}$ increases. Eventually as $S(\rho_A) \rightarrow 0$, x_A and $x_{\bar{A}}$ pinch off from one another altogether.

So Van Raamsdonk's second case illustrates the existence of a correlation between entanglement entropy and distance. Indeed, this correlation is a natural one given that as the entanglement entropy decreases, the mutual information $I(A:\bar{A})$ also decreases, and (as Cao et al. 2017 also note) the mutual information has several features (positivity, symmetry, the obeying of a triangle inequality) that make it particularly apt to be tracking distance relations. But what reason is there to think this correlation is due to the *emergence of the facts about distance from the facts about entanglement*? Van Raamsdonk's case for the correlation and corresponding images comes simply from an appeal to the Ryu–Takayanagi conjecture (2010, 3). And so, to evaluate whether this connection comes from the fact that the spatiotemporal relations are emerging from entanglement on the boundary (in support of my Option (1)), we must go back to see what justifies the Ryu–Takayanagi area law in the first place.

We will return to that momentarily, but first let us consider Van Raamsdonk's first case in which we see a pair of disconnected AdS spacetimes becoming spatiotemporally connected, 'glued', using quantum entanglement. First, Van Raamsdonk asks us to consider a pair of CFTs, which we may call CFT_1 and CFT_2, each defined on $S^3 \times \mathbb{R}$. In a product state, their combined quantum state may be written as a tensor product, $\rho_{AB} = \rho_A \otimes \rho_B$. In this case, there is no entanglement between the degrees of freedom of the two CFTs. And so their individual entanglement entropies are identically zero. Here, CFT_1 and CFT_2 describe two separable systems, and so, Van Raamsdonk argues, the dual AdS spacetimes AdS_1 and AdS_2 are disconnected.[11]

[11] One might already question this inference, at least insofar as it is supposed to generalize beyond the case of AdS/CFT, where we may assume each CFT gives a complete description of an AdS spacetime.

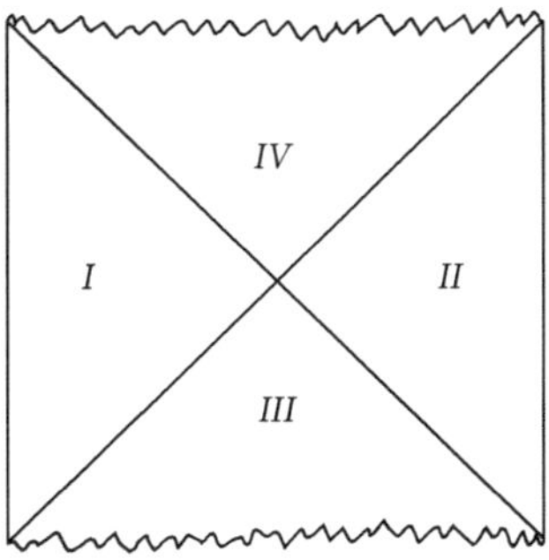

Fig. 4.7 Eternal AdS black hole spacetime

However, what if we take two such unentangled CFTs and combine them into a quantum superposition? Van Raamsdonk asks us to consider the following superposition of product states:

$$|\psi\rangle = \sum_i e^{\frac{-\beta E_i}{2}} |E_i\rangle \otimes |E_i\rangle . \tag{4.9}$$

This state, the Hawking–Hartle state, has been shown by Maldacena (2003) to correspond to a classically connected spacetime, the eternal AdS black hole spacetime, which can be represented by the above Penrose diagram (Fig. 4.7): Here, diagonal lines represent the light cone structure. Regions I and II correspond to the duals of the reduced states for CFT_1 and CFT_2, respectively. Jagged lines represent the past and future singularities.

Van Raamsdonk concludes how remarkable it is that such a complicated quantum state can give rise to a spacetime with such a simple geometric description in classical terms:

> ...we have a remarkable conclusion: the state $|\psi\rangle$ which clearly represented a quantum superposition of disconnected spacetimes may also be identified with a classically connected spacetime. In this example, *classical connectivity arises by entangling the degrees of freedom in the two components.* (2010, 2)

But is this the correct order of explanation? In the eternal AdS black hole spacetime, the causal futures of the two asymptotic regions AdS_1 and AdS_2 intersect past the event horizon. This is their joint causal future, Region IV. As Van Raamsdonk notes, it follows from the AdS/CFT correspondence that if we take a measurement

Outside of the AdS/CFT context, we have no problem conceiving separable states of quantum systems where the component states ρ_A and ρ_B inhabit a common spacetime.

on a state described by CFT_1, the dual of the result of that measurement will eventually enter region IV. This is the causal future of the dual to CFT_2. This suggests that measurement of CFT_1 affects the state of CFT_2, and supports the claim that the two are entangled. Thus, it seems it is the spatiotemporal connection between AdS_1 and AdS_2 that partially grounds the entanglement between the dual CFT_1 and CFT_2. But if this is correct, then Van Raamsdonk's discussion of the eternal AdS black hole seems most suggestive of the idea that the order of explanation goes in the opposite direction than what he wants. It is the spacetime structure that explains why it is that the measurement in the dual CFT_1 has an effect on the measurement in the dual CFT_2, and so explains the entanglement of the two component states. If this is right, then it is not that the geometric facts emerge from the quantum facts, but rather that the former constrain the latter.

Return to the four metaphysical options presented in Section 4.4.

1. Facts about the metric emerge from fundamental facts about entanglement entropy.
2. Facts about the entanglement entropy emerge from fundamental facts about the metric.
3. Facts about the metric and entanglement entropy are equally fundamental.
4. Facts about the metric and entanglement entropy are neither fundamental, but both emerge from some more fundamental, as yet undiscovered, framework.

My claim is that Van Raamsdonk's first case does not support Option (1) over its rivals, even if it does support the existence of an explanatory connection between facts about entanglement entropy and facts about the metric. The explanation of the entanglement relations in this case presupposes facts about the metric.

4.6 Back to Ryu–Takayanagi and Black Hole Thermodynamics

Let's try to seek out arguments for the emergence claim that are more promising by returning to the Ryu–Takayanagi conjecture and seeing in the first place why this conjecture turns out to work.

Recall again that Ryu and Takayanagi propose the following area law:

$$S(\rho_A) = \frac{\text{area of } \gamma_A}{4G_N^{(d+1)}}, \tag{4.10}$$

where $G_N^{(d+1)}$ is Newton's constant for the relevant $(d+1)$-dimensional space. It's easiest to understand the case for this area law in the context of a specific physical scenario. Let's consider the simplest case of $d = 2$. Here we are considering an AdS_3 gravity theory that is dual to a (1 + 1)-dimensional CFT. For an AdS_3 spacetime with radius of curvature R, the dual CFT has central charge $c = 3R/2G_N^{(3)}$.

Now note that the metric of the AdS_3 spacetime is:

$$ds^2 = R^2\left(-\cosh\rho^2 dt^2 + d\rho^2 + \sinh\rho^2 d\theta^2\right). \tag{4.11}$$

The proper distances diverge as ρ goes to infinity. Although this divergence is not problematic on its own, for there to be a finite boundary on which the dual CFT will yield values for physical quantities that do not diverge, we impose a cutoff ρ_0 and restrict the space of interest to $\rho \leq \rho_0$. In the dual CFT, this corresponds to imposing a UV cutoff where there is a minimal distance between field values of a, the lattice spacing (cf. Susskind and Witten 1998). We can then think of the CFT as living on the boundary of AdS_3 where $\rho = \rho_0$. This makes the $(1 + 1)$-dimensional spacetime for the CFT a cylinder (t, θ) with radius ρ_0.

Ryu and Takayanagi consider the static case where the CFT lies on a single time slice t (see Fig. 4.8). As mentioned, Hubeny et al. (2007) later extended this argument to more general regions to give a covariant result. To give the main idea of how this is established, I will focus on the static case and closely follow the 2006 discussion. Take a particular time-slice of the cylinder and divide it into a region A and its complement, such that A corresponds to $0 \leq \theta \leq \frac{2\pi l}{L}$, where l is the length of A and L is the length of the total boundary. The minimal surface γ_A then will be the geodesic with fixed t that connects the points $\theta = 0$ and $\theta = \frac{2\pi l}{L}$. This will be a geodesic running through the AdS_3 spacetime.

Ryu and Takayangi compute the length of the geodesic from the metric as $1 + 2\sinh^2\rho_0 \sin^2\frac{\pi l}{L}$. And then using this for the area of γ_A, they use the proposed area law (4.10) to get an expression for $S(\rho_A)$. The neat thing is that the result given by

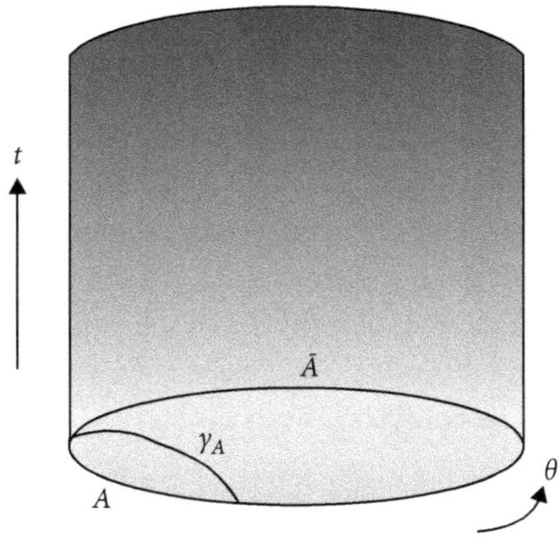

Fig. 4.8 Static case

the area law is confirmed by independent work on the entanglement entropy of many body (1 + 1)-dimensional CFTs. As Ryu and Takayanagi note, Calabrese and Cardy (2004) showed (generalizing results of Holzhey et al. 1994) that for (1 + 1)-dimensional CFTs, the entanglement entropy of a quantum many-body system (with periodic boundary conditions) is given by:

$$S\left(\rho_A\right) = \frac{c}{3}\log\left[\frac{L}{\pi a}\sin\left(\frac{\pi l}{L}\right)\right], \tag{4.12}$$

where again L is the length of the total system, l is the length of A, a is the lattice spacing, and c is the CFT's central charge. Recalling that our CFT has central charge $c = 3R/2G_N$, we then recover:

$$S\left(\rho_A\right) = \frac{c}{3}\log\left(e^{\rho_0}\sin\frac{\pi l}{L}\right), \tag{4.13}$$

assuming $e^{\rho_0} \gg 1$.

The form of argument Ryu and Takayanagi use may be summarized as follows. Using the AdS_3 metric, we can compute the area of a minimal surface connecting the edges of region A through the bulk. If we then assume the area law, this allows us to calculate a certain formula for the entanglement entropy $S(\rho_A)$ in the language of conformal field theory. This turns out to be the correct formula for $S(\rho_A)$ within a (1 + 1)-dimensional CFT, given independent results. This thus confirms the area law. Ryu and Takayanagi develop similar arguments for more complicated setups in higher dimensions (2006a, 2006b).

So we get confirmation of Ryu and Takayanagi's conjecture of an area law, but this does raise the question of why such an area law obtains. What grounds the connection between $S(\rho_A)$ and the area of γ_A? Discussions of Ryu and Takayanagi's paper often take their result to support the emergence of spacetime from an underlying quantum theory without gravitational degrees of freedom, but although their arguments certainly confirm an entanglement-distance link, they don't provide the explanation of why this link should hold, and as such cannot support Option (1) over its three rivals. Here I would suggest it is helpful if we think about the Bekenstein–Hawking law, which inspired Ryu and Takayanagi's conjecture. Bombelli et al. (1986) develop an illuminating physical argument that provides an explanation of why the Bekenstein–Hawking relationship holds. As it works by explicating Bekenstein–Hawking entropy as a form of entanglement entropy, it can help us to better understand the metaphysical implications of the Ryu–Takayanagi conjecture, which essentially takes the Bekenstein–Hawking result and generalizes it outside of the black hole context.[12]

[12] It should be noted that arguments based on the claim that black hole entropy is entanglement entropy do not provide the only path to explaining the Bekenstein–Hawking formula. For example, one also finds arguments that one can arrive at the relationship just by counting states. To see how

The core idea introduced in Sorkin (1983) and developed in Bombelli et al. (1986) is to treat the source of black hole entropy as related to its entanglement entropy, and find the entanglement entropy of the black hole by considering the entanglement entropy of the region outside it.[13] This is illustrated using a simple model, where the states of the total region consist of the states of a massless scalar field in its vacuum state. To calculate the entanglement entropy, we can represent the field as a lattice of harmonic oscillators with density $1/\ell^3$. This allows us to represent the entanglement entropy as a sum over eigenvalues λ_i for an operator Λ, using the previously stated relation:

$$S(\rho) = \sum_i \lambda_i \log \frac{1}{\lambda_i} \tag{4.14}$$

This operator Λ encodes the correlations between points in the lattice and importantly depends on a division of points in the lattice into those internal and external to the black hole. See Bombelli et al. (1986) for the details and Sorkin (1983) for an overview, but the core idea is that assuming a finite number of points in the lattice, the main contribution to the entanglement entropy will come from the values of Λ closer to the horizon (cf. Carlip 2009). And so, given how Λ is constructed, "the leading term in S will be proportional to $\frac{A}{\ell^2}$ where A is the area of the surface ('horizon') separating H^{int} from H^{ext}" (Sorkin 1983, 3).[14] The entropy of a black hole thus scales with the area of its event horizon. This is because the sum in (4.14) is dominated by terms describing the states of the field near the boundary, and thus the size of the boundary between the regions inside and outside of the black hole is imposing a constraint on the leading contribution to $S(\rho_{ext})$.

We can see why this is by considering that any physically acceptable state of the field will be a Hadamard state.[15] As noted in Unruh and Wald (2017, 2), the leading order behavior for Hadamard states Ψ of a free scalar field ϕ at two points x_1 and x_2 is:

$$\langle \Psi | \phi(x_1)\, \phi(x_2) | \Psi \rangle \sim \frac{U(x_1, x_2)}{(2\pi)^2 \sigma(x_1, x_2)}. \tag{4.15}$$

This is the two-point function of the field, where $U(x_1, x_2)$ is some smooth function and $\sigma(x_1, x_2)$ is the squared geodesic distance between the two points. From (4.15),

this is done in the string theory context, see e.g. Strominger and Vafa (1996). However, the form of argument I will describe and closely related arguments presented e.g. by Srednicki (1993) are the most salient if one wants an explanation that can be carried over to the Ryu–Takayanagi area law which concerns entanglement entropy.

[13] Sorkin introduced the idea of entanglement entropy in this 1983 paper.

[14] H^{int} and H^{ext} are the two spacelike hypersurfaces we are considering inside and outside of the black hole.

[15] Wald (1994, chapters 2–3) argues that the requirement that states be Hadamard ensures that the stress-energy tensor is defined.

it is easily seen that the two-point function diverges as x_1 and x_2 approach one another. And so, as the operator Λ encodes the correlations between points inside and outside of the event horizon, the largest contributions to (4.14) will come from those right at the event horizon.

Bombelli et al.'s derivation of an area law for the entanglement entropy of a black hole, at least the contribution to this entropy given by a massless scalar field, justifies the posited area law of Bekenstein and Hawking. And so we can use this justification to ask about the justification for the area law of Ryu and Takayanagi.

4.7 No Support for Emergence

It is clear from the above discussion of black hole entropy that there is no antecedent reason to view the area of the event horizon as constructed out of or emerging from the entanglement of the reduced quantum state ρ_{BH}. The most natural way to interpret the argument is that the horizon area is providing a constraint on the black hole entropy. It is not emerging from the entropy.[16] Insofar as the Ryu–Takayanagi conjecture is providing a generalization of the black hole area law, in that case too there doesn't seem to be any reason to view the area of the minimal surface γ_A as constructed out of the entanglement entropy. It is more natural again to view the area of the minimal surface as providing a constraint on the entanglement entropy. And so, the burden of proof is on one who wants to interpret the area law of Ryu and Takayanagi in a very different way metaphysically from the area law it was aimed at generalizing. This point carries over to the work following Ryu and Takayanagi (e.g. Hubeny et al. 2007, Bao et al. 2019, and so on). It is worth noting that Ryu and Takayanagi do not speculate about metaphysics in their original paper. Moreover, in a longer work (2006b) that followed it, expanding on its implications, they present their results not as motivating the emergence of spacetime but rather in an instrumental way: since it is often mathematically very difficult to compute the entanglement entropy from the quantum field theory side of things, one can make use of this proposed area law to get a good approximation of the entanglement entropy (2006b, 2–3).

One might argue that although the Ryu–Takayanagi conjecture was *inspired by* the Sorkin–Bombelli explanation of the area law of Bekenstein and Hawking, and although there are important analogies between black hole entropy and entanglement entropy on a CFT boundary, this does not entail that there is

[16] This is the most natural interpretation of 't Hooft's (1993) holographic proposal for black hole entropy as well, as he argues that the black hole entropy is maximal when it reaches $S_{max} = \frac{1}{4}A$. "[The entropy] is saturated when one black hole has the largest possible size that still fits inside our area" (5).

an absolute alignment between the two cases.[17] Although in the case of the Bekenstein–Hawking law, facts about spatiotemporal distances do explain the entanglement entropy, one might argue that in the case of the AdS/CFT correspondence, facts about entanglement entropy are fundamental and instead explain the facts about spatiotemporal distance. This is a coherent possibility. However, what I am trying to argue is that a more natural attitude is to view the Ryu–Takayanagi result as supporting the generalization of the Sorkin–Bombelli framework outside of the context it was initially intended to describe. And thus that in general, entanglement entropy is constrained by interface (event horizon or minimal surface) area. Interface area doesn't emerge from entanglement entropy in either case.

To make the case for Option (1), one needs to point to arguments in which entanglement entropies are plausibly being used to explain metrical features, without thereby assuming them. One may immediately think of the influential paper by Jacobson (1995), which derives the Einstein Field Equation from facts about entanglement entropy. But Jacobson himself relies on the Bekenstein–Hawking relationship (and the Sorkin–Bombelli explanation of it) in his argument. So this doesn't give us what we are looking for. Work by Padmanabhan (2010a,b) comes closer, as it attempts to provide an explanation of the spacetime metric using a more basic statistical mechanical framework for understanding facts about entanglement entropy. Padmanabhan explicitly frames his work as a part of a project in emergent gravity:

> In this paradigm, one considers spacetime (described by the metric, curvature etc.) as a physical system analogous to a gas or a fluid (described by density, velocity etc.). The fact that either physical system (spacetime or gas) exhibits thermal phenomena shows there *must* exist microstructure in either system... We do not yet know what are the correct microscopic degrees of freedom of the spacetime, but the horizon thermodynamics provides a clue... (2010a, 3)

Padmanabhan (2010b) shows how one can use the equipartition law $\Delta E = (1/2)(\Delta n)k_B T$ to derive the existence of these fundamental microscopic degrees of freedom and their relationship to the spacetime metric such that:

[17] There is one clear disanalogy between the cases that is worth mentioning. This is that the area that appears in the Bekenstein–Hawking relation is the area of the boundary between the two subsystems (interior and exterior of the black hole's event horizon), but the area that appears in the Ryu–Takayanagi relation is not the area of the boundary between the two subsystems (A and its complement). In the latter case, the area in question is the area of the minimal surface through the bulk that ends on the boundary between A and its complement. This disanalogy, however, does not undermine the above argument, as the boundary between A and its complement itself plays a role in constraining the minimal surface area.

$$\Delta n = \frac{\Delta A}{l_P^2}, \tag{4.16}$$

where l_P is the Planck length. The idea seems to be that the facts about entanglement entropy motivate the postulation of these fundamental degrees of freedom, which in turn ground the structure of gravity and emergent spacetime. This is an intriguing possibility, but at this stage, it is worth noting, given we don't have any idea what these microscopic degrees of freedom are, or a strong motivation for believing in them, how much more speculative these arguments are than the arguments we saw above that show how facts about the entanglement between components of quantum systems are constrained by facts about the interface surface between these components. Then again, as Carlip (private communication) points out, the latter kind of arguments have had a 'big head start'.

Let's finally return to the case for area laws outside of the AdS/CFT context, the argument we find in the results summarized in Eisert et al. (2010) and followed up on by e.g. Cao et al. (2017). This route into the entanglement-distance link also fails to provide support for the view that the facts about spatiotemporal distance emerge from facts about entanglement entropy. As we saw, the way area laws were derived for quantum many body systems relied on an initial assumption of local interactions. The idea is that when influences are only able to act locally, then the limits on correlation and hence entanglement will be restricted by the region's surface area. In this case again, there doesn't seem to be support for the emergence hypothesis. Again, what seems supported, indeed presupposed, by the whole discussion is that the area of the boundary is constraining the entanglement of the two component regions (for Eisert et al., these are I and its complement), by constraining the quantity of direct interactions. It is certainly much more exciting and radical to view the entanglement-distance link as resulting from a fact about emergence—that the spatiotemporal manifold is not fundamental, but ultimately results from deeper, quantum information theoretic facts about quantum entanglement. But the way these results are motivated does not support this radical view. In the framework in which the results are articulated, the spatiotemporal facts are fundamental. They then constrain facts about interactions, which then in turn constrain the facts about entanglement entropy.

4.8 Conclusion

What I have hoped to show is that there is strong support both from within and without the AdS/CFT context for the existence of an important link between entanglement entropy and spatiotemporal distance. However, these results do not lend obvious support for the speculation that the dynamical spacetime of general relativity is emergent from facts about quantum entanglement; at least, this is not supported by work to date. One has to interpret results like those of Ryu and Takayanagi in a way that departs significantly from the intuitions that make them comprehensible to have them support the claim that spacetime emerges from the facts of quantum entanglement.

This isn't to say that one cannot use the results of Ryu and Takayanagi and others to support the claim that the metric emerges from facts about quantum entanglement. In a recent book, Carroll suggests we do just that. He says:

> ... following our reverse-engineering philosophy, we can *define* the 'area' of a collection of degrees of freedom to be proportional to its entanglement entropy. In fact, we can assert this for every possible subset of degrees of freedom, assigning areas to every surface we can imagine drawing within our network... If the degrees of freedom are highly entangled, we *define* them to be nearby... A metric on space has emerged from the entanglement structure of the quantum state. (2019, 284)

Of course, this is a move one can make, and certainly the attendant emergence hypothesis is one we should keep on the table, especially as it seems supported by other routes into developing a quantum theory of gravity (cf. Wüthrich 2017). However, the goal of this chapter has been to make clear that this is not what area laws most directly support, when we pay attention to how they have typically been justified in physics.[18]

References

Audenaert, K., J. Eisert, M. B. Plenio, and R. F. Werner (2002). Entanglement properties of the harmonic chain. *Physical Review A 66*(4), 042327.

Balasubramanian, V., J. de Boer, M. P. Heller, B. Czech, and B. D. Chowdhury (2014). A hole-ographic spacetime. *Physical Review D 89*, 086004.

Bao, N., C. Cao, S. Fischetti, and C. Keeler (2019). Towards bulk metric reconstruction from extremal area variations. *Classical and Quantum Gravity 36*(18), 185002.

Bekenstein, J. D. (1973). Black holes and entropy. *Physical Review D 7*(8), 2333.

Bombelli, L., R. K. Koul, J. Lee, and R. D. Sorkin (1986). Quantum source of entropy for black holes. *Physical Review D 34*(2), 373–383.

[18] Thanks to Steve Carlip, Veronika Hubeny, Nick Huggett, and an anonymous referee for comments and discussions that substantially improved this chapter.

Butterfield, J. (2021). On dualities and equivalences between physical theories. In C. Wüthrich, B. Le Bihan, and N. Huggett (Eds.), *Philosophy Beyond Spacetime*. Oxford: Oxford University Press. This volume.

Calabrese, P. and J. Cardy (2004). Entanglement entropy and quantum field theory. *Journal of Statistical Mechanics: Theory and Experiment 2004*(06), P06002.

Cao, C., S. M. Carroll, and S. Michalakis (2017). Space from Hilbert space: Recovering geometry from bulk entanglement. *Physical Review D 95*(2), 024031.

Carlip, S. (2009). Black hole thermodynamics and statistical mechanics. In E. Papantonopoulos (Ed.), *Physics of Black Holes*, pp. 89–123. Springer.

Carlip, S. (2012). Challenges for emergent gravity. *Studies in History and Philosophy of Modern Physics 46*(2), 200–208.

Carroll, S. (2019). *Something Deeply Hidden: Quantum Worlds and the Emergence of Spacetime*. New York: Dutton.

Crowther, K. (2014). *Appearing out of nowhere: The emergence of spacetime in quantum gravity*. Ph.D. thesis, University of Sydney.

De Haro, S. (2017). Dualities and emergent gravity: Gauge/gravity duality. *Studies in History and Philosophy of Modern Physics 59*, 109–125.

De Haro, S. (2020). Spacetime and physical equivalence. In N. Huggett, K. Matsubara, and C. Wüthrich (Eds.), *Beyond Spacetime: The Foundations of Quantum Gravity*, pp. 257–283. Cambridge: Cambridge University Press.

Dieks, D., J. van Dongen, and S. de Haro (2015). Emergence in holographic scenarios for gravity. *Studies in History and Philosophy of Modern Physics 52*, 203–216.

Eisert, J., M. Cramer, and M. B. Plenio (2010). Colloquium: Area laws for the entanglement entropy. *Reviews of Modern Physics 82*(1), 277.

Harlow, D. (2020). What black holes have taught us about quantum gravity. In N. Huggett, K. Matsubara, and C. Wüthrich (Eds.), *Beyond Spacetime: The Foundations of Quantum Gravity*, pp. 105–114. Cambridge: Cambridge University Press.

Hawking, S. W. (1975). Particle creation by black holes. *Communications in Mathematical Physics 43*(3), 199–220.

Headrick, M., R. C. Myers, and J. Wien (2014). Holographic holes and differential entropy. *Journal of High Energy Physics 2014*(10), 149.

Holzhey, C., F. Larsen, and F. Wilczek (1994). Geometric and renormalized entropy in conformal field theory. *Nuclear Physics B 424*(3), 443–467.

Horowitz, G. T. (2005). Spacetime in string theory. *New Journal of Physics 7*(1), 201.

Hubeny, V. E. (2014). Covariant residual entropy. *Journal of High Energy Physics 2014*(9), 156.

Hubeny, V. E., M. Rangamani, and T. Takayanagi (2007). A covariant holographic entanglement entropy proposal. *Journal of High Energy Physics 2007*(07), 062.

Huggett, N. and C. Wüthrich (2013). Emergent spacetime and empirical (in)coherence. *Studies in History and Philosophy of Modern Physics 44*(3), 276–285.

Huggett, N. and C. Wüthrich (forthcoming). *Out of Nowhere: The Emergence of Spacetime in Quantum Theories of Gravity*. Oxford: Oxford University Press.

Jacobson, T. (1995). Thermodynamics of spacetime: The Einstein equation of state. *Physical Review Letters 75*(7), 1260.

Jaksland, R. (2020). Entanglement as the world-making relation: Distance from entanglement. *Synthese*. https://doi.org/10.1007/s11229-020-02671-7.

Maldacena, J. (1999). The large-n limit of superconformal field theories and supergravity. *International Journal of Theoretical Physics 38*(4), 1113–1133.

Maldacena, J. (2003). Eternal black holes in anti-de sitter. *Journal of High Energy Physics 2003*(04), 021.

Myers, R. C., J. Rao, and S. Sugishita (2014). Holographic holes in higher dimensions. *Journal of High Energy Physics 2014*(6), 44.

Nielsen, M. A. and I. L. Chuang (2016). *Quantum Computation and Quantum Information*. Cambridge: Cambridge University Press.

Oriti, D. (2021). Levels of spacetime emergence in quantum gravity. In C. Wüthrich, B. Le Bihan, and N. Huggett (Eds.), *Philosophy Beyond Spacetime*. Oxford: Oxford University Press. This volume.

Padmanabhan, T. (2010a). Finite entanglement entropy from the zero-point area of spacetime. *Physical Review D 82*(12), 124025.

Padmanabhan, T. (2010b). Surface density of spacetime degrees of freedom from equipartition law in theories of gravity. *Physical Review D 81*(12), 124040.

Preskill, J. (1998). Lecture notes for physics 229: Quantum information and computation. *California Institute of Technology 16*. http://theory.caltech.edu/~preskill/ph229/.

Rangamani, M. and T. Takayanagi (2017). *Holographic Entanglement Entropy*. Springer.

Read, J. (2016). The interpretation of string-theoretic dualities. *Foundations of Physics 46*(2), 209–235.

Rickles, D. (2013). AdS/CFT duality and the emergence of spacetime. *Studies in History and Philosophy of Modern Physics 44*(3), 312–320.

Ryu, S. and T. Takayanagi (2006a). Aspects of holographic entanglement entropy. *Journal of High Energy Physics 2006*(08), 045.

Ryu, S. and T. Takayanagi (2006b). Holographic derivation of entanglement entropy from the anti-de sitter space/conformal field theory correspondence. *Physical Review Letters 96*(18), 181602.

Sorkin, R. D. (1983). On the entropy of the vacuum outside a horizon. In B. Bertolli, F. de Felice, and A. Pascolini (Eds.), *Tenth International Conference on General Relativity and Gravitation (Padova, 4–9 July, 1983), Contributed Papers*, Volume 2, pp. 734–736.

Srednicki, M. (1993). Entropy and area. *Physical Review Letters 71*(5), 666–669.

Strominger, A. and C. Vafa (1996). Microscopic origin of the Bekenstein–Hawking entropy. *Physics Letters B 379*(1–4), 99–104.

Susskind, L. and E. Witten (1998). The holographic bound in anti–de sitter space. arXiv hep-th/9805114.

't Hooft, G. (1993). Dimensional reduction in quantum gravity. In *Conference on Highlights of Particle and Condensed Matter Physics (SALAMFEST)*, pp. 284–296. World Scientific.

Teh, N. J. (2013). Holography and emergence. *Studies in History and Philosophy of Modern Physics 44*(3), 300–311.

Unruh, W. G. and R. M. Wald (2017). Information loss. *Reports on Progress in Physics 80*(9), 092002.

Van Raamsdonk, M. (2010). Building up spacetime with quantum entanglement. *General Relativity and Gravitation 42*(10), 2323–2329.

Wald, R. M. (1994). *Quantum Field Theory in Curved Spacetime and Black Hole Thermodynamics*. University of Chicago Press.

Witten, E. (1998). Anti de sitter space and holography. *Advances in Theoretical and Mathematical Physics 2*(2), 253–291.

Wüthrich, C. (2017). Raiders of the lost spacetime. In D. Lehmkuhl, G. Schiemann, and E. Scholz (Eds.), *Towards a Theory of Spacetime Theories*, pp. 297–335. Basel: Birkhäuser.

5
Taking Up Superspace
The Spacetime Setting for Supersymmetric Field Theory

Tushar Menon

Abstract

Supersymmetry (SUSY) is a proposed symmetry between bosons and fermions. The structure of the space of SUSY generators is such that the distinction between internal and spacetime symmetries is blurred. As a result, there are two viable candidates for the spacetime setting for a flat supersymmetric field theory—Minkowski spacetime and superspace, an extension of Minkowski spacetime to include (at least) four new dimensions, coordinatized by 'super-numbers', i.e. numbers with nontrivial commutation properties. This chapter argues for two theses: first, that one standard set of arguments, related to universality of symmetry behaviour, that motivate a particular choice of spacetime structure in familiar spacetime theories motivates the choice of superspace as the appropriate spacetime for SUSY field theories; and second, that the metaphysical utility of the concept of spacetime requires more than just the satisfaction of this universality condition; in supersymmetric theories, the spacetime concept is not as useful as in special relativity.

5.1 Introduction

Supersymmetry (SUSY) is a proposed symmetry between quantum fields of integer spin (bosons) and quantum fields of half-integer spin (fermions). SUSY is a bizarre and brilliant idea, and it deserves to be scrutinized by philosophers. In this chapter, we will merely dip our toes into its deep waters. The motivation for the project of looking at SUSY from a philosophical perspective is not as radical as it might appear at first. Philosophical intuitions are heavily influenced by, and tied to, particular frameworks of physical theories. Exploring the logical space of such possibilities has value for the light it sheds on the constraints imposed by our current theorising. As Friedman (2001) argues, physical theories are embedded in a package of background formal and conceptual assumptions, which can be

Tushar Menon, *Taking Up Superspace: The Spacetime Setting for Supersymmetric Field Theory* In: *Philosophy Beyond Spacetime: Implications from Quantum Gravity.* Edited by: Christian Wüthrich, Baptiste Le Bihan, and Nick Huggett, Oxford University Press. DOI: 10.1093/oso/9780198844143.003.0005

hard to see explicitly, or appear as matters of necessity when they are seen. SUSY places current spacetime and particle physics in a broader logical landscape, revealing hidden assumptions and contingencies. In this regard, its epistemic value is independent of whether or not SUSY is realized in nature and this is especially clear when seen in light of Friedman's own interest in 'a genuine expansion of our space of intellectual possibilities'.

Further pragmatic motivation for studying SUSY comes from the role that it plays in string theory, one of our prime candidates for a quantum theory of gravity. Any theory which purports to describe our world needs to have the conceptual resources to describe fermions. The only way that we know of to incorporate fermions into a string theory is through SUSY (for more on this, see e.g. Becker et al. 2006, chapters 4–5, Zwiebach 2004, chapter 14). So any thesis regarding spacetime in a string theory that hopes to model our actual world must include reference to the role that SUSY plays. In this chapter, I choose to focus only on SUSY, divorcing it from the context of string theory, but it is useful to bear this motivation in mind throughout.

In this chapter, I attempt to identify the appropriate spacetime setting for a supersymmetric field theory. Whatever the motivations, I hope to make clear that SUSY introduces a new dimension (several, actually) to the discussion, and places under the microscope our motivations for being interested in spacetime structure in the first place.

Constructing an appropriate spacetime for a supersymmetric theory is tantamount to taking a stance on the viability of a spatiotemporal interpretation of *superspace*[1]—an extension of four-dimensional Minkowski spacetime to include (at least) four new dimensions, coordinatized by mathematical objects known as *supernumbers*. These objects are, in one significant way, quite different from real or complex numbers—some of them have the property that their order of multiplication makes a difference; in mathematical terms, they are said to have *non-trivial commutation properties*.

This chapter argues for two theses: first, that some standard arguments that motivate a particular choice of spacetime structure in familiar spacetime theories also motivate the choice of superspace as the appropriate spacetime for SUSY field theories; and second, that the broader metaphysical utility of the concept of spacetime requires more than just the satisfaction of this universality condition. In short, superspace is spatiotemporal, but in an attenuated sense; in the terminology of Read and Menon (2019), superspace is a theoretical, but not operational, spacetime.

[1] This is not to be confused with the term 'superspace' (or any associated terms like 'mini-superspace'), as used in the context of geometrodynamics in e.g. Isenberg and Nester (1980); Misner et al. (1973). They use the term to refer to a mathematical space of spatial 3-metrics.

This chapter is structured as follows. In §5.2, I examine how dynamical symmetry considerations factor into arguments concerning spacetime structure. In §5.3, I briefly introduce SUSY by analogy with constructions familiar from standard spacetime theories. In §5.4, I discuss problems and considerations analogous to those brought up in §5.2, this time in the context of SUSY, and show that all of the arguments presented in that section mandate a choice of superspace, rather than Minkowski spacetime, as the appropriate spacetime setting for SUSY. Finally, in that section, I discuss, in more detail, motivations for being interested in spacetime structure, and whether the identification of superspace as the appropriate spacetime setting for a certain class of SUSY field theories is theoretically valuable.

5.2 Arriving at Spacetime Structure

Our only means of epistemic access to spacetime structure (whatever its associated metaphysics) is through the behaviour of dynamical fields.[2] As a contingent fact about a particular dynamical field, it might be the case that certain configurations exhibit a behaviour that allows for the (possibly approximate) measurement of proper time according to some metric along intervals of a worldline. As a further contingent fact, it might be the case that a number of different matter fields can be brought into configurations which measure the same intervals of proper time along their worldlines. If it is the case that *all* of the matter fields to which we have access are such that intervals of proper time along their worldlines are seen to correspond to the same metric, then, on purely practical grounds, it makes sense to see this structure as codifying some universal features of the dynamics of these fields. In particular, granting this 'universality' supposition, we can then make claims about certain types of behaviour (e.g. the measurement of length intervals, claims about signalling and causality) in a manner which is neutral on further idiosyncratic details of the dynamics; this is what is sometimes taken to be characteristic of (some of the) *kinematical* structure in dynamical theories. The methodological importance of spacetime symmetries is precisely in their ability to encode some of these kinematical facts. This much is uncontroversial. Further disputes that may arise over claims about what explains this feature of dynamical fields constitute the dynamical–geometrical debate, more details of which can be found in, e.g., Brown (2005); Brown and Read (forthcoming).

[2] I use the term 'dynamical fields' to include test bodies, i.e. matter fields that do not themselves act as sources in the Einstein field equations. Thus deducing metric structure from, for example, conformal and projective structure à la Ehlers, Pirani, and Schild (2012) from the behaviour of test particles (and light rays) still qualifies as using dynamical fields.

It is important to be clear about the use to which the spacetime concept is to be put. What I have described above is a fairly minimal requirement, but one might be interested in even less—instead of encoding facts about the behaviour of certain contingent matter agglomerations (e.g. rods and clocks), we might think of spacetime structure as encoding nothing more than certain geometrical facts common to all dynamical fields. On this view, all there is to being a spacetime symmetry is to be a transformation such that when applied to *any* dynamical field, solutions are mapped to solutions. In Read and Menon (2019), the associated invariant structure was referred to as *theoretical spacetime*, to contrast it with *operational spacetime* which deals with the behaviour of matter configurations like rods and clocks; call such matter configurations *surveyors*.

- **Theoretical spacetime**: That geometrical structure whose automorphisms are the (common) symmetries of the dynamical fields.
- **Operational spacetime**: That structure which is correctly surveyed by physical surveyors built from matter fields.

One might require even more than what I described above—that the appellation 'spacetime' refers to some structure that reproduces our first-personal experience of a world through which we 'move' in some sense; call this *phenomenal spacetime*. In this chapter, my interest is in the relationship between theoretical and operational spacetime; phenomenal spacetime will not be discussed. We will see that the theoretical spacetime concept is useful for metaphysical theorizing in virtue not only of the universality condition, but also other contingent facts about certain types of dynamical fields. Equipped with this distinction, let us now review some standard proposals concerning spacetime structure and determine whether they refer to theoretical or operational spacetime.

5.2.1 Earman's Principle

Earman presents his famous prescription—make your dynamical and spacetime symmetries (as defined below) coincide—henceforth referred to as *Earman's principle* in the second chapter of *World Enough and Spacetime* (1989). In order for this dictum to make sense, we need to be clear about what spacetime and dynamical symmetries are. In what follows, I briefly discuss two proposals for determining which symmetries count as spacetime symmetries. In §5.2.1.1, I describe Earman's own proposal and its drawbacks. I propose my own criteria for identifying spacetime symmetries in §5.2.1.2.

5.2.1.1 Earman on Absolute Objects

On a standard way of understanding the structural setup of a spacetime theory, a theory is identified with (or, more loosely, associated with) a set of models.[3] Models can be thought of as set-theoretic entities whose mathematical structure is isomorphic, in an appropriate sense, to (some of) the structure of the world. A model $\mathcal{M}$ of a spacetime theory is an ordered tuple of the form $\langle M, A^i, P^i \rangle$ where M is the set of independent variables of our theory (for a classical field theory, this is taken to be a four-dimensional smooth, Hausdorff, metrizable manifold).

The A^is represent the *absolute (geometric) objects* which characterize the spacetime. An absolute object can be thought of as a geometric object which is identical across all models. Consider, for example, *Newtonian spacetime*, which privileges a certain inertial frame corresponding to a standard of absolute space. Such a frame would be identified using a privileged timelike vector field in this formalism.

The P^is are geometric objects which represent the dynamical elements of the theory—the matter fields and force fields which are subject to dynamical equations. For a classical field theory, these will be maps from M to some appropriate mathematical space in which the fields take their values. The models thus specified form the set of *kinematically possible models* (KPM). The dynamics can be specified by stipulating which of these maps are nomologically allowed. A theory is then simply the collection of models whose fields obey these dynamical constraints, the *dynamically possible models* (DPM).[4]

On this setup, it is easy to distinguish between dynamical and spacetime symmetries. The former are the transformations to the P^is under which the dynamical equations retain their form. These transformations are actually carried out by pushing forward the P^is along (a subset of)[5] diffeomorphisms on the base manifold. The condition that the dynamical equations retain their form is the requirement that these diffeomorphisms map DPM onto DPM. A diffeomorphism

[3] This view of scientific theories is known as the *semantic view*, and was introduced by Suppes (1961).

[4] This construction glosses over an important technical point about whether the absolute objects remain unchanged across all kinematically possible models (KPM) or merely across all dynamically possible models (DPM). I take Anderson's (1967) terminology of absolute objects to refer to the latter, while Pooley's (2017) 'fixed fields' refer to the former. For the purpose of this chapter, the Andersonian conception of absolute objects suffices.

[5] There is a subtlety here which is discussed at length by Pooley (forthcoming, 128, 134 footnote 41). When dealing with coordinate-free formulations, *everything* is, by construction, covariant under diffeomorphisms (a property known as general covariance, and distinct from diffeomorphism invariance as characterized by Pooley 2017). All this means is that if $\mathcal{M}$ is a model of a theory, then so is the tuple obtained by pushing forward *all* the objects defined on the manifold. But these two models need not be identical. If they are not, then the subgroup of diffeomorphisms which leaves each of them invariant will be different for each model. But they will be isomorphic as groups. If the models are of the form $\mathcal{M} = \langle M, A^i, P^i \rangle$, then the abstract group which is constrained to act only on the A^i to leave the models unchanged will be the spacetime symmetry group, while the group which acts on the P^i to map it onto another model will be the dynamical symmetry group.

which, when the A^is are pushed forward along it, leaves the A^is invariant is known as a *spacetime symmetry.* In other words, spacetime symmetries leave invariant the absolute objects of a theory. One might then consider the base manifold together with its absolute objects to constitute the *background structure* of the theory. The spacetime symmetries are then the automorphisms of this structure. The method for picking out absolute objects comes from some independent considerations (usually extra-theoretic, metaphysical, or methodological). At least, this is the case on the coordinate-independent view presented here. Friedman himself follows this route in arriving at the symmetry group of a theory:

> [T]he symmetry group . . . of a theory is the largest subgroup of . . . the group of automorphisms . . . leaving certain of the geometrical objects of the theory— . . . the absolute objects of the theory—invariant. (Friedman 1983, 42)

The problem with this setup is that it assumes that we have some independent handle on, at the very least, the sorts of things that could qualify as absolute objects (or fixed fields). This prescription then allows us to narrow down the options for this structure.

The examples presented in Earman's own treatment constitute a highly restricted class of spacetime theories. In particular, the fields and particles are assumed to have no degrees of freedom other than positions and momenta. Incorporating, say, a single $U(1)$–symmetric dynamical gauge field, in addition to the point particles, would, according to Earman's principle, require the spacetime symmetries to include that group. An obvious response to this worry is to somehow naturally restrict the class of dynamical symmetries to which Earman's principle is to apply to the so-called *external* dynamical symmetries.

5.2.1.2 Spacetime Symmetries from External Symmetries

Earman takes both parties to have agreed on the sort of thing that constitutes a spatiotemporal relation; their disagreement is over what grounds these relations. Earman himself does not offer any further clarification on this matter: spatiotemporal relations are things like relative distances, angles, and time intervals. Implicit in this, therefore, is some universality constraint which would explain our interest in the kinematical structure of spacetime. In this subsection, I make that intuition precise.

A state of a system is specified by some collection of values of variables. Some of these variables are *independent*, which means we are free to choose their values, and others *dependent*, which means that they are determined by certain functions on the independent variables which we had chosen. This distinction is one of convenience, and it is most common for the set of independent variables in a field theory to be identified with spacetime points. Since this is the very structure we are trying to get an independent handle on, we cannot presuppose it.

Consider the state of a complex-valued classical Lorentz-invariant field. It is specified by an uncountable number of tuples of complex numbers. We are interested in dividing this parameter space into 'internal' and 'external' parameter spaces. This is where our antecedent knowledge of the symmetry structure of the theory comes in. Given the dynamical symmetries of this theory (say, $U(1)$ and $SO(1,3)$), the parameter space splits naturally into a two-dimensional parameter space invariant under $U(1)$ rotations and a four-dimensional parameter space invariant under $SO(1,3)$ rotations. This split is the basis of the internal/external distinction. If we now consider a number of other Lorentz-invariant fields, their parameter spaces will also split. Further, they will all have the four-dimensional $SO(1,3)$-invariant parameter space in common. The symmetry group of this common parameter space is what I will refer to as *external dynamical symmetries*. Note that, in the case of a theory with just one field, although the natural split still exists, there is no principled reason to refer to one parameter space rather than the other as the 'external' space.

We can use Earman's principle to structure spacetime appropriately. For example, in Newtonian theory, force-free particles execute straight-line trajectories in spacetime. The external dynamical symmetry group is the Galilean group (the internal symmetry group is the singleton $\{e\}$, the identity). Earman's principle tells us that the spacetime symmetry group must, therefore, also be the Galilean group. In other words, the subset of the diffeomorphism group which induces push-forwards on the absolute objects of the theory is the Galilean group.

It is important to bear in mind that our identification of external dynamical symmetries with spacetime symmetries is *extensional*. We assume that absolute objects exist, and that they have some associated symmetry group. We then use Earman's principle as a guide for how to use dynamical symmetries to ascertain this structure. The dynamical and spacetime symmetry transformations themselves act on distinct objects, and it is a methodological convention that they are co-extensional. And, using the terminology introduced in Read and Menon (2019), the resulting spacetime is a *theoretical* spacetime.

We can think of Earman's principle as mandating that we use the most minimal structure that allows us to encode universal dynamical facts. As such, it can be seen as an Occamist norm on our theorizing. But the question that now arises is why we ought to restrict our attention to external dynamical symmetries. In the following sections, I canvass three interpretations of spacetime structure that account for this restriction.

5.2.2 The Geometrical Approach

There appears to be a simple answer to the question posed above—we restrict to external symmetries just because they *are* the symmetries of the underlying spacetime geometry. In other words, we assume that spacetime, whatever it is, can be represented by a smooth manifold with some geometrical structure on it. This geometrical structure is privileged in the sense that it is pervasive—all dynamical fields adhere to its directives. This is the central assumption of the so-called *geometrical approach* to spacetime as advocated by, e.g., Friedman (1983), Belot (2000), and Maudlin (2012). The mechanism by which this is achieved is disputed; indeed, whether such a mechanism (in the quotidian physical sense of it being some dynamical process) is even required is up for debate. But, whatever the reason, on this view it is the case that matter fields are compelled, by the geometry, to evolve in such a way as to survey it. Ultimately, absent a detailed dynamical account for why this is the case, this view is justified by appeal to some form of abductive reasoning.

On this approach, we are in a good epistemic situation. All we need to do is pick a suitably stable matter configuration, and use it to measure distances, angles, and time intervals. If we corroborate these readings with other suitably chosen matter configurations of different materials, this eventually gives us a sufficiently accurate picture of the underlying geometry of spacetime, using Earman's principle. External dynamical symmetries are thus, by definition, the symmetries of matter fields which match the symmetries of the underlying geometry (i.e. the spacetime symmetries) because they preserve the 'pervasive' geometrical structure. This is, of course, just a necessary, not sufficient condition for matter to 'survey' this geometry, i.e. for configurations to read off intervals of the time parameter associated with worldlines embedded in that geometry; this is a specific case of the observation presented in Read and Menon (2019) that theoretical spacetime structure does not, in general, guarantee operational spacetime structure. However, in a large class of physically relevant spacetimes (e.g. Minkowski, Schwarzschild, FLRW) the standard matter that we use to construct clocks and rods is such that theoretical spacetime structure *does*, in fact, determine the metric structure that they read off.

Thus the geometrical approach and Earman's principle together form a package; indeed, the latter requires the former in order to be meaningfully formulated. In §5.2.3, I drop the assumption that some independent element of reality (like a geometry) compels matter to behave in a specific way.

5.2.3 The Dynamical Approach

The dynamical approach answers the above question of why we restrict ourselves to external symmetries in a different way. Rather than presupposing that an independent element of the world, represented by a geometry, is the ultimate explanans of the spatiotemporal (i.e. pervasive or universal) behaviour of matter fields, the proponent of the dynamical approach asserts that the dynamics itself is the ultimate explanans. Of course, this does lead to an unanswered question about why there is such a large symmetry coincidence across all known matter field theories—on the dynamical approach, this is just an unexplained brute fact about the world.[6]

For a Brown-style proponent of the dynamical approach, a spacetime symmetry is connected to the behaviour of chronometric matter fields, i.e. matter fields which can read off intervals of proper time along a trajectory as given by the metric field g_{ab}; therefore as a prescription for identifying operational spacetime, the dynamical approach only works with solutions to the Einstein field equations in which theoretical spacetime structure guarantees operational spacetime structure (as in the aforementioned examples of Minkowski, Schwarzschild, and FLRW solutions). The dynamical approach is, however, perfectly consistent with theories in which this link does not exist.

In summary, on the dynamical approach, (theoretical) spacetime structure is wholly determined by the behaviour of matter fields. Thus, a version of Earman's principle holds trivially—all external dynamical symmetries will be spacetime symmetries, because the latter can only be defined as external dynamical symmetries.

5.2.4 Spacetime Functionalism

The dynamical approach technically does not commit its adherent to a particular view on *spacetime*; Brown himself is only interested in physical geometry; whether the structure thus identified deserves the appellation 'spacetime' is additional. Knox's spacetime functionalism (as presented in Knox 2013, 2019) is one way of executing this additional step (note, however, that spacetime functionalism is compatible with the geometrical approach as well). For Knox, spacetime is any structure in a physical theory that plays a role in characterizing inertial structure; she defines inertial frames as:

1. Inertial frames are frames with respect to which force free bodies move with constant velocities.
2. The laws of physics take the same form (a particularly simple one) in all inertial frames.
3. All bodies and physical laws pick out the same equivalence class of inertial frames (universality). (Knox 2013, 348)

[6] For a deeper discussion of this fact, see Brown (2005); Read et al. (2018).

Knox builds into her notion of inertial structure, through the second criterion, the transformation behaviour of matter field equations that the proponent of the geometrical view builds into its characterization of the metric. At this point, then, it is clear that spacetime functionalism guarantees theoretical spacetime. But we know, from Knox's own writings, that her inertial frame spacetime functionalism is motivated by a desire to capture operational spacetime as well. Consider, for example, the following passage:

> [C]onsidering the inertial structure provides a shortcut that allows us to glean the empirical consequences of a theory without going into the messy details of our various measuring devices. (Knox 2013, 347)

This makes sense given the intended domain of Knox's programme—classical spacetime theories like Newtonian Gravitation, and special and (some solutions of) general relativity. In such scenarios—scenarios in which, as will be discussed in §5.2.5, the clock hypothesis is satisfied—theoretical spacetime structure is enough to determine operational spacetime structure.

5.2.5 How to Do Things with Spacetime

It seems intuitively obvious why one might be interested in operational spacetime; an account of the reasons that measuring devices behave in the way that they do tells us a great deal about the dynamical structure of the matter fields in the world—after all, measuring devices form part of the structure that allows for the testing of hypotheses. Another reason (which I will not address further in this chapter) is to overcome the problem of *empirical incoherence*, a term first used in the context of quantum mechanics by Barrett (1999), of quantum gravity theories. Briefly, the worry is that many (perhaps all) of our best guesses as theories of quantum gravity seem to indicate that operational spacetime disappears at sufficiently small scales because of quantum effects. All of our empirical science, however, seems to be predicated on the idea that scientific data about the world is recorded by spatiotemporally located devices. Therefore, if some particular theory of quantum gravity without spacetime is true, then the truth of that theory would undermine any empirically motivated reasons we might have for believing in it. Huggett and Wüthrich's response (2013) is to point out that there is no requirement that spacetime be fundamental; an emergent account of spacetime would suffice as long as such an account can be robustly provided. Indeed, this is arguably an arena in which spacetime functionalism is of significant value.

Our reasons for being interested in theoretical spacetime are less obvious, beyond its being a necessary component of an account of operational spacetime.

Here, I offer one reason for our interest in theoretical spacetime for its own sake—its role in our metaphysical theorizing about the world. Several debates in metaphysics turn on accounts of space and time—causation, causality, identity, modality, and the nature of time itself, to name a few. But getting bogged down in the complicated dynamical details of matter fields in the actual world would preclude a great deal of this theorizing. This is why so-called kinematical facts about space and time play such a vital role—they abstract away from conceptually irrelevant contingent characteristics of particular dynamical theories, and allow for general claims to be advanced about the world. Take causation and causality, for example. It is now taken for granted that faster-than-light communication is prohibited by relativity theory. But it is extremely uncommon (I know of no examples) for a metaphysical text to demonstrate this fact by, for example, solving Maxwell's equations in the geometrical-optical limit in all relevant spacetimes and demonstrating the physical significance of the light-cone structure of each tangent spaces of these spacetimes. Far more common (and sensible) is to begin by just assuming that this structure is, in fact, a feature of spacetime.

It will turn out that the three procedures outlined in this section—Earman's, Brown's, and Knox's—all agree that superspace is spatiotemporal. However, the utility of this fact is dampened by the severance of the link between theoretical spacetime structure and operational spacetime structure. This link, in the sorts of spacetime theories considered so far, is provided by the *weak clock hypothesis*:

- *Weak clock hypothesis*: There exist configurations of matter whose dynamics is such that they can be used, in inertial as well as suitably gently accelerated states of motion, to record intervals of proper time, as determined by some metric, along their worldlines.[7]

In special relativity, the clock hypothesis is defined with respect to the Minkowski metric, η_{ab}; in general relativity, with respect to the metric g_{ab} of the Einstein field equations. The local symmetry coincidence given by theoretical spacetime structure guarantees the Lorentzian signature of the metric, at least for matter fields described by the (classical approximation of the) standard model. Given local symmetry coincidence of the dynamical laws and the geometry, the minimal condition one needs to satisfy to be able to have epistemic access to operational spacetime *approximately* is that certain matter configurations are *boostable*, i.e. they are just as good at reading off intervals of proper time along a metric before and after an active Lorentz boost, even if not *during* the

[7] This restriction to 'suitably gently accelerated states of motion' is important. To expect the clock hypothesis to hold for *any* state of motion is far too restrictive; every physical clock has some breaking point.

boost. This is a strictly weaker condition than satisfaction of the weak clock hypothesis.

The weak clock hypothesis is a field-dependent statement—it makes reference to certain field configurations. However, it is standard to think of the clock hypothesis as being field-neutral, i.e. the choice of the metric with respect to which the hypothesis is defined is chosen in such a way as to be independent of the choice of matter field.[8] Let us therefore refine the above definition to a *strong clock hypothesis* (this is what is often referred to simply as 'the clock hypothesis'; see e.g. Fletcher 2013, Maudlin 2012):

- *Strong clock hypothesis*: For each kind of matter, there exist appropriate configurations whose dynamics is such that they can be used to record, in an inertial or suitably gently accelerated state of motion, intervals of proper time along their worldlines, as determined by some metric. Further, this is the metric with respect to which clocks built out of other matter fields will record proper times along their worldlines.

Satisfaction of the strong clock hypothesis then guarantees that theoretical spacetime structure determines operational spacetime structure. The strong clock hypothesis is what ensures the universality of a class of spatiotemporally oriented metaphysical theses. We will see, in §5.4, that the kinematical structure of superspace is not as friendly to such metaphysical theorizing as Minkowski spacetime is, because of the failure of the strong clock hypothesis.

5.3 Supersymmetric Field Theory and Superspace

In order to extend the arguments presented in the previous section, we will need a small number of new formal tools. The material presented in this chapter is sufficient for just this purpose; it is not intended to constitute anything like an introduction to SUSY for philosophers. In particular, I make a number of important technical restrictions. For example, I consider only *superclassical* fields—these are fields that have the algebraic properties of bosonic and fermionic fields, but are themselves classical field-theoretic and not quantum field-theoretic objects[9]—that are symmetric under rigid (i.e. constant across the space of independent variables)

[8] For the proponent of the dynamical approach, the physical utility of this move is a brute unexplained fact; for the proponent of the geometrical approach, this is explained by the constraining power of the metric.

[9] A quantum field is standardly interpreted as a spacetime-indexed operator-valued distribution.

and non-extended (i.e. there is only one irreducible spin representation of the SUSY generators) SUSY transformations. The physics presented in this section is based on Wess and Bagger (1992) and Weinberg (2000); the mathematics on DeWitt (1992), Buchbinder and Kuzenko (1998), and Rogers (2007).

SUSY can be approached from a number of directions; in this chapter, our trajectory is mapped out by our interest in SUSY as a spacetime symmetry. I will therefore mostly ignore its quantum field-theoretic roots, and present it as a generalization of a spacetime symmetry (in the sense of §5.2). This section is structured as follows. In §5.3.1, I define, in the most general possible terms, SUSY transformations and then discuss how SUSY, a transformation between particle types, can be given an interpretation as a spacetime symmetry. In §5.3.2, I introduce the necessary mathematical concepts to set up superspace. Finally, I present, in §5.3.3, a simple supersymmetric field theory.

5.3.1 What Is Supersymmetry?

SUSY is a collective term for a set of transformations between *bosons* (which typically represent the carriers of force, such as photons, w-particles, or gravitons) and *fermions* (which typically represent the quanta of matter, such as electrons or quarks) that leaves the Lagrangian density associated with a particular theory invariant (up to a possible surface term). This is the first symmetry of its kind in physics—until now, all symmetry transformations have been such that bosons have been taken to other bosons and fermions to other fermions, leaving a force-matter distinction invariant.

A natural expectation, on learning the definition of SUSY, is that an even number of applications of a SUSY transformation returns the original particle type (for example, a boson would switch to a fermion then back to a possibly distinct boson). This is indeed what happens, but the process has a surprising consequence—the transformed boson is displaced in spacetime compared to the original. It is this novel mixing of symmetry types that gives SUSY such philosophical promise.

One can learn everything about the symmetry structure of a physical theory by studying the groups associated with the symmetry transformations under consideration. So the study of rotations in Euclidean space, for example, reduces to the study of the *orthogonal group*. Physicists often use a trick to simplify the mathematics—they focus on infinitesimal versions of the transformations, and study the associated *algebras*. For most theories, this reproduces most of the interesting structural features of the theory at a fraction of the mathematical cost. We can, therefore, learn a good deal about SUSY (whose dynamical symmetry group is the *super-Poincaré* group) by studying its algebra, known as the *super-Poincaré* algebra.[10]

Infinitesimal SUSY transformations are represented by an operator, Q_α: applying Q_α to a boson, ϕ, turns it into a fermion, ψ. The process of 'acting on an object', e.g. turning a boson into a fermion, is enacted mathematically by finding a *representation* of the algebra element, Q_α. The algebraic structure of the set of such operators is encoded in what is known as the *anti-commutator*: for each pair of operators Q_α and Q_β, the sum of the results of acting with Q_α first and Q_β second, and the other way round. It encodes the information about the repeated application of the transformation to which we referred above. We find, on working through the mathematics, that the anti-commutator of two of these generators is, completely unexpectedly, proportional to the generator of spacetime translations.[11] We therefore conclude that there is something putatively spatiotemporal, at least at the infinitesimal level, about SUSY. In other words, a purely *internal* transformation (i.e. one which explicitly does not involve changing the position or orientation of an object in spacetime) between kinds of matter has somehow led to a spacetime displacement.

It is natural to think of a symmetry transformation as being something that does not change the identity of the object on which it acts. The fact that the dynamics does not see this change is an indication that this is a good inference. A rotation of, say, a plate might change its configuration, but it does not change the fact that the rotated object is still a plate, albeit in a (possibly indistinguishable) different state. So how do we make sense of a *symmetry* transformation between a boson and a fermion, when it is clear that both of these objects are *distinct*?[12] This is a two-stage process.

The first stage is to realize that, if SUSY is, in fact, a symmetry of the actual world, then the current state of the world is one in which the symmetry is *broken*. We come across broken symmetries all the time, even in classical physics. The laws of mechanics are translation-invariant, but that does not mean that if I suddenly move myself twenty feet to the right, I won't notice a change. What has happened

[10] Technically, the super-Poincaré algebra is a *superalgebra*, corresponding to the super-Poincaré group, which is a *super-Lie group*. Their role in SUSY theories is analogous (with certain technical caveats explored in Rogers 2007) to the role of a symmetry algebra and its corresponding group in ordinary field theory.

[11] The explicit form of the super-Poincaré algbera relations is:

$$\{Q_\alpha, Q^\dagger_\beta\} = (\sigma^\mu)_{\alpha\beta} P_\mu$$

$$[Q_\alpha, P_\mu] = 0$$

$$[Q_\alpha, M_{\mu\nu}] = (\sigma_{\mu\nu})_\alpha^{\ \beta} Q_\beta$$

[12] Even referring to bosons as objects is not uncontroversial. See Saunders (2003).

is that my state is no longer symmetric under all the transformations under which the laws are symmetric.

The second stage is to consider the regime of unbroken SUSY, and think of bosons and fermions as being components of a more general type of field. This is known as a *superfield*, and is the focus of §5.3.3. In order to define a superfield, we need to be able to talk about supernumbers and superspace.

5.3.2 Supernumbers and Superspace

Supersymmetric laws are invariant under a transformation of a bosonic field to a fermionic field and vice versa. To simplify things a little, we work with so-called superclassical fields. Quantum operators, as elements of an algebra that is not necessarily commutative, are the sorts of things that can commute or anti-commute. All classical fields, on the other hand, commute, so they are represented as maps from spacetime to some commuting space (usually this is also a manifold and vector space; standard examples are n-tuples of real or complex numbers). To construct a superclassical field, all we do is widen the scope of the target space of the maps to include spaces of anticommuting numbers (and tuples thereof). We do not interpret these as operator-valued distributions; instead, we see that they take their values in spaces of (possibly tuples of) number-like objects known as *supernumbers*.

5.3.2.1 Supernumbers

In the introduction, we characterized supernumbers as generalizations of real and complex numbers in such a way as to have non-trivial commutation properties. An explicit construction of these objects, as elements of an infinite *Grassmann algebra*, can be found in Buchbinder and Kuzenko (1998). For the purposes of this chapter, we can think of supernumbers as elements of a space of objects (of uncountable cardinality) which splits into two subspaces. The first consists of (a set of objects isomorphic as an algebra to the) real numbers, and other objects that commute with each other (and the real numbers); the second consists of objects that *anticommute*. If ξ_1 and ξ_2 are supernumbers, then if $\xi_1 \cdot \xi_2 = -\xi_2 \cdot \xi_1$, but $\xi_1 \neq 0 \neq \xi_2$, these supernumbers anticommute. In many respects that are significant to the physics, they behave like ordinary scalars—they can be added, subtracted, multiplied, and divided. These spaces are, respectively, the space of real commuting supernumbers, denoted by $\mathbb{R}_c$, and the set of all real anticommuting supernumbers, $\mathbb{R}_a$. In what follows, $x^a \in \mathbb{R}_c$ and $\theta^a \in \mathbb{R}_a$. It is notationally convenient to define θ (and its complex conjugate, $\bar{\theta}$) as belonging to the *complex* anticommuting supernumbers—these are merely ordered pairs of real anticommuting numbers.

We now have the resources to define the classical analogues of fermionic fields as maps from some spacetime manifold (more on this in §5.3.3) to the space of anticommuting supernumbers (and tuples composed therefrom), and the classical analogue of bosonic fields as maps into the space of commuting supernumbers and tuples composed therefrom (this is a space that includes the real numbers as well).

5.3.2.2 Superspace

We would like to write down classical field equations which are symmetric under interchange of commuting and anticommuting fields, and impose this symmetry in the standard field-theoretic way, by considering representations of the algebra of infinitesimal generators. Since we are dealing with fields which transform into one another, and given that the two fields have very different algebraic properties, we need a new kind of algebra. It turns out that what we need are infinitesimal anticommuting generators, infinitesimal commuting generators, and a rule for how they interact. This is the information contained in the super-Poincaré algebra.

Generalizing the spacetime manifold is the crucial step that allows us to construe superspace as being spatiotemporal; this generalized manifold is the titular superspace. It is standard to construct a manifold by beginning with a set of points, call it M, defining maps from these points into the real numbers (i.e. coordinatizations), and then specifying the structure on M by specifying the allowed coordinatizations of M. This is the Kleinian approach to geometry (for more details, see e.g. Anandan 1980, Wallace 2019), and modulo limited success with trying to define this structure nominalistically (see Arntzenius 2012, chapter 8), it is the only way in which we can construct a smooth manifold.

A *supermanifold* is constructed by analogy with an ordinary manifold, using supernumber-valued coordinatizations.[13] It is the product space of real commuting supernumbers and real anticommuting supernumbers (under the DeWitt topology and supersmooth structure; see Rogers 2007). Formally, the definition of a superspace, denoted by $\mathbb{R}^{p|q}$, is

$$\mathbb{R}^{p|q} := \mathbb{R}^p_c \times \mathbb{R}^q_a = \{z^M = (x^1, x^2, ...x^p, \theta^1, \theta^2, ...\theta^q),\ x^m \in \mathbb{R}^p_c,\ \theta^\mu \in \mathbb{R}^q_a\}.$$

A point in such a space now has $p + q$ coordinates, p of which are commuting supernumbers, q of which are anticommuting supernumbers. An infinitesimal translation in the superspace coordinates can be easily defined:[14] The transformations are

$$x'^a = x^a - i\varepsilon\sigma^a\bar{\theta} + i\theta\sigma^a\bar{\varepsilon}, \tag{1}$$

[13] There are a number of thorny technical issues that come up. For example, because of the nilpotent elements, there is no natural Hausdorff topology on the space of supernumbers in the way that there is on the real numbers. So the generalization to a supermanifold loses some of the physically desirable qualities of ordinary manifolds. For more details on this construction, see Rogers (2007, chapters 2–6).

[14] A discussion of this can be found in Buchbinder and Kuzenko (1998, 158).

and

$$\theta'^{\alpha} = \theta^{\alpha} + \varepsilon^{\alpha}, \tag{2}$$

where the α index runs from 1 to q, and σ^a are the Pauli spin matrices. A similar expression to (2) holds for the complex conjugate coordinate $\bar{\theta}$. With these transformations, along with the standard Lorentz transformations,[15] it is easy to find an expression for the analogue of a Minkowski interval for superspace, a function that is left invariant by these transformations. This two-point function, defined below, is the *super-Minkowski interval.*

$$w^a(z_1, z_2) = (x_2 - x_1)^a + i\theta_\alpha \sigma^a(\bar{\theta}_\beta - \bar{\theta}_\alpha) - i(\theta_\alpha - \theta_\beta)\sigma^a \bar{\theta}_\alpha. \tag{3}$$

where z_1 and z_2 are coordinates in superspace. The super-Minkowski interval, arrived at here using coordinates, can be thought of as expressing the action of an absolute geometric object (in the sense of §5.2); call it the *super-Minkowski metric.*

5.3.3 Superfields

A real classical scalar field $\phi(x)$ can be represented as:

$$\phi(x) : \mathbb{R}^4 \to \mathbb{R}_c. \tag{4}$$

This idea can be generalized to incorporate a set of both commuting as well as anticommuting fields. Each element of a set of dynamical fields $\{\phi^i\}$ can be represented as

$$\phi^i(x) : \mathbb{R}^4 \to \mathbb{R}^{p|q}, \tag{5}$$

where p is the number of commuting field components and q the number of anticommuting ones.[16]

Finally, one might generalize this procedure to produce an object, call it a *generalized superfield*, Φ, that transforms as a generalized rotation in superspace, much like a Lorentz boost can be thought of as a rotation in spacetime:

$$\Phi^i : \mathbb{R}^{4|4} \to \mathbb{R}^{p|q}. \tag{6}$$

[15] It is worth noting here that there is no 'super' analogue of a Lorentz transformation; the super-Poincaré group is the semi-direct product of the ordinary Lorentz group and the super-translation (super)group.

[16] We temporarily gloss over the fact that we need to do some more work in order to ensure that certain essential features of our fields, like smoothness, translate appropriately into this formalism—the short answer is that they do. A precise definition of what it is for a function to be 'supersmooth', for example, can be found in Rogers (2007).

With the setup presented above, it is possible to combine the hitherto separate types of field—bosonic and fermionic—into a single field, Φ. At this stage, there is nothing radical about this move. Expressing fields in terms of combinations of other fields is a common move in QFT—it is the first step in constructing a Lagrangian for a field theory with a spontaneously broken symmetry, for example. Consider, now, a mapping from this superspace (supermanifold) into the superspace of real commuting supernumbers:

$$V : \mathbb{R}^{4|4} \to \mathbb{R}_c.$$

On Taylor expanding this superfield in its anticommuting coordinates, we recover all the expected bosonic and fermionic fields (for simplicity, I have returned to using θ_1 and θ_2 instead of θ_α and $\bar{\theta}_\alpha$):

$$V(x, \theta_1, \theta_2) = A(x) + B(x)\theta_1 + C(x)\theta_2 + D(x)\theta_1\theta_2 \tag{7}$$

where $A(x)$ and $D(x)$ are commuting (bosonic) fields, and $B(x)$ and $C(x)$ are anticommuting (fermionic) fields.

Notice how, by incorporating the supernumber-valued coordinates into the argument of the function V, we have been able to express both fermionic and bosonic fields (we do not, as yet, know that they are bosonic/fermionic—it is their transformation behaviour under the Poincaré group/algebra which determines that) as functions of commuting coordinates, weighted by the appropriate commuting or anticommuting factor. The mathematical advantage of this superspace setup is clear—the equations of motion associated with these fields are manifestly supersymmetric. In the next section, we discuss the physical significance of interpreting superspace as a theoretical spacetime.

5.4 On the Spatiotemporality of Superspace

We now have all the pieces in play to decide on the spacetime structure of a classical flat supersymmetric field theory. In §5.4.1 I revisit the three approaches discussed in §5.2. I then discuss the issue of chronometry in §5.4.2 and see how the link between theoretical and operational spacetime is severed in superspace.

5.4.1 The Three Approaches to Spacetime

5.4.1.1 Earman's Principle

Using the generalized mathematical notions introduced in §5.3, we can set up a supersymmetric field theory in a model-theoretic way that allows us to articulate Earman's principle. The models of a flat superclassical field theory are tuples of the form $\mathcal{M} = \langle SM, A^i, P^i \rangle$, where SM is a supermanifold. The dynamical symmetries are elements of the super-Poincaré group, by construction. Since this is the largest common symmetry group, by the arguments presented in §5.2, this qualifies as the analogue of the external symmetry group as well. So Earman's principle mandates that these be the symmetries of the absolute objects as well; the 'spacetime' symmetry group. This is why one of the absolute objects characteristic of this 'spacetime' is a generalized tensor superfield that defines the super-Minkowski interval on the supermanifold.

Contrast this with the alternative approach that maintains that the absolute object be only the Minkowski metric tensor, η_{ab}. The symmetries of this object are just the Poincaré transformations. But the dynamical symmetry group continues to be the super-Poincaré group. This leads to an unacceptable mismatch, and a violation of Earman's principle. Therefore, the proponent of this principle has to accept that superspace (equipped with the relevant geometrical object that specifies the super-Minkowski interval) is a theoretical spacetime.

5.4.1.2 The Dynamical Approach

On the dynamical approach, we begin with the fields themselves, and ask what their (largest) common dynamical symmetry group is. There is no underlying commitment to an independently existing spacetime metric, to which we might only have epistemic access through the behaviour of matter fields; the spacetime metric is itself nothing more than a codification of the symmetry properties of matter fields. Therefore we are forced to conclude that, for a universally super-Poincaré-invariant theory, only superspace equipped with the super-Minkowski metric function could be spatiotemporal; the ordinary Minkowski metric simply does not encode all aspects of the universal behaviour of matter fields.

5.4.1.3 Spacetime Functionalism

Equipped with the generalization of the Minkowski interval, the super-Minkowski interval defined in (7), it is possible to construct the superspace analogue of inertial frames, called *super-inertial* frames.[17] Recall that a crucial ingredient of Knox's definition of inertial frames was a universality requirement. This requirement can only be satisfied by taking the relevant relativity principle to be the one expressed by the statement that the largest common external symmetry group is the super-Poincaré group—this way we avoid making reference, in our definition of inertial frames, to velocities and force-free motion. We can then run the generalization of Knox's argument in superspace—what it is to be a spatiotemporal degree of freedom is to play a role in determining the super-inertial structure of a theory.

The set of dynamical symmetry transformations common to all superfields is, by construction, the *super-Poincaré group*—this is an example of a 'super-Lie

[17] For more on these frames, see Buchbinder and Kuzenko (1998, chapter 2).

group'. Super-Lie groups are a generalization of Lie groups; whereas the latter are groups which are also smooth manifolds, the former are groups which are also superspaces (i.e. generalizations of manifolds with anti-commuting supernumber-valued coordinates). Recall that a theoretical spacetime is a geometrical structure determined by constructing geometrical objects that are invariant under transformations from the dynamical symmetry group common to all fields in question. By construction, therefore, superspace is a theoretical spacetime.

Knox's prescription identifies as 'inertial' those frames in which the laws take their simplest form. In the case under consideration here, these are those frames in which the dynamical laws governing the superfields are expressed in coordinates in which the super-Minkowski interval, determined by the super-Minkowski metric, is invariant. Thus, the super-Minkowski metric plays the role of picking out super-inertial frames—it therefore qualifies as a theoretical spacetime.

To see if the super-Minkowski metric qualifies as an operational spacetime, however, we need to assume that one can construct the appropriate generalization of rods and clocks. Let us call these objects *super-surveyors*. An example of a super-surveyor would be the analogue of a light clock, but one in which the oscillating material is governed by the supersymmetric version of Maxwell's equations, the so-called *super-Maxwell* equations. But, as is discussed in more detail in the next section, it is far from clear that the extra SUSY dimensions can be surveyed by such matter field configurations.

5.4.2 Chronometry in Superspace

In discussions on the foundations of relativity theory, surveyors are generally treated as epistemologically convenient matter configurations whose behaviour is such that they allow us to measure lengths and durations as given by the underlying spacetime metric. Nothing in the foundations of relativity theory, therefore, requires surveyors to be primitive.

5.4.2.1 What Is It to Be Spatial?

Minkowski famously made a proclamation in 1909 about the fate of space and time as distinct concepts:

> Henceforth space by itself, and time by itself, are doomed to fade away into mere shadows, and only a kind of union of the two will preserve an independent reality. (Minkowski 1909)

There is a sense in which this is true—the symmetry group of special relativity includes transformations on coordinates that 'rotate' spatial coordinates into temporal ones and vice versa. However, this does not mean that, physically, our best theories do not distinguish between spatial and temporal directions—they had better, given the starkly different ways in which spatial and temporal evolutions are treated (cf. e.g. Skow 2007). Callender makes the important observation that "although relativity banishes 'time' and 'space', there is nevertheless a sense in which it is committed at its very core to a difference between timelike and spacelike directions" (Callender 2017, 124).

What Callender means by this is that, even in Minkowski spacetime, there is a physical distinction between spacelike and timelike directions. Moreover, this distinction survives a Poincaré transformation—there is no such transformation that will change a spacelike direction into a timelike one or vice versa. And this fact is encoded in the structure of the Minkowski metric tensor, η_{ab}.

In its most general form, a *relativity principle* asserts the equivalence of a description of physics in one frame of reference with that in another suitably related frame of reference. The Galilean relativity principle, for example, asserts that the form of dynamical equations is the same in all inertial frames, i.e. frames that are moving at a constant velocity with respect to each other. A relativity principle brings along with it a notion of invariance. The value of this invariant quantity is put in by hand; recall that in special relativity, for example, it follows from another postulate, the *light postulate*. For Galilean relativity, the invariant notion is of spatial distance—the length of a rod before and after any Galilean transformation is unchanged. What is distinctive about the special relativistic relativity principle is that the invariant quantity is a *speed*, not a spatial length. This is, of course, a trivial consequence of assuming the light postulate, which asserts that the two-way speed of light is isotropic and independent of the speed of the source. Without the light postulate, the existence of *some* invariant quantity is guaranteed by the relativity principle. The light postulate adds three physically salient elements to this—(i) that the invariant quantity is a *speed*, (ii) the actual value of the invariant speed, and (iii) that some physical matter fields (i.e. Maxwell fields) exist with that speed of propagation.

Now, assuming the relativity principle (and that spacetime is homogeneous and space is isotropic), the Lorentz transformations can be derived, given the light postulate. A different value for the speed of light would lead to a different group of transformations between inertial frames—they would still be groups of Lorentz transformations, but each group would be parametrized by a different value of c. Thus, the light postulate and the relativity principle together determine the (inner-product) structure of the tangent spaces.

There is a problem with attempting to generalize this move to superspace—the absence of an empirical principle analogous to the light postulate. It is possible to define the tangent space at a point in superspace, in effect by specifying algebraically the space of tangent vectors along anticommuting dimensions—

for details, see Buchbinder and Kuzenko (1998, 406). When one postulates that the transformations between super-inertial frames are super-Poincaré transformations, but does not say anything further, the parameter that determines the tangent space ('inner-product') structure for the anticommuting dimensions is left unspecified.

If we believe that our surveying devices have to be built out of SUSY-matter (and that seems reasonable; how else could they survey the superspace geometry?), then it is possible that this matter will not read off intervals of the super interval, even if, with respect to the Minkowski subspace of superspace, the strong clock hypothesis may be satisfied. The reason is the following: all our superfields are (by construction) super-Poincaré invariant—but they might, nonetheless, still survey distinct supermetrics because they do not agree on the structure of the anticommuting component of the tangent space to superspace. Thus, super-surveyors constructed in this way will not be guaranteed to give us operational access to spacetime.

I should stress that I do not intend to suggest that supersymmetric matter will somehow propagate faster than light on its Minkowski dimensions; the structure of the Poincaré subgroup ensures that it will not. But generic features of superspace geometry will be accessible to super matter if and only if it satisfies a generalization of the clock hypothesis, the *super-surveyor hypothesis*:

- *Super-surveyor hypothesis*: There exist matter configurations that read off generalized metric intervals along the trajectories of particles in any physically suitable state of motion. Further, this is the generalized metric with respect to which super-surveyors built out of other matter fields will record intervals along their worldlines.

We have good reason to suspect the clock hypothesis is approximately satisfied; this evidence comes from observation, not from any specific prediction from within general relativity. In other words, the clock hypothesis cannot be derived from within general relativity.[18] In the absence of empirical or extra-theoretical reasons to believe in the satisfaction of the super-surveyor hypothesis, we cannot make claims about super-surveyors in the way that we do about surveyors in special or general relativity. Our only alternative is to look for a mathematical indication from within supersymmetric field theory that a super-surveyor hypothesis could be satisfied. Unfortunately, our attempts to do so in special and general relativity are not promising. In what follows, I provide two short heuristic arguments to cast doubt on the possibility of such a proof in a superclassical Yang–Mills theory based on disanalogies with Maxwellian electromagnetism in special relativity.

[18] This was one of the morals of Menon et al. (forthcoming).

First, as discussed above, the clock hypothesis in special relativity relied on the validity of the light postulate, which asserts a claim about the independence of the speed of light from the speed of the source of that light. If one looks for a justification from within Maxwellian electromagnetism, one eventually hits some different phenomenological postulate. For example, that the speed of light is constant in all inertial frames follows from the statement that the magnetic permeability, μ_0, of the classical vacuum and the electric permitivity of the classical vacuum, ε_0, are the same in all inertial frames. We can thus arrive at the light postulate in reverse, from assuming the frame-independence of the values of these constants. But those facts too are merely postulated, and ultimately only justified by appeal to empirical observation, the likes of which we have no access to in superspace.

The surveyors we have considered so far have all been implicitly modelled as Langevin clocks—perfect reflectors with some oscillating material bouncing between them—which emphasize that the crucial link between clocks and chronometry is through the notion of an oscillation. Oscillations, whether of some material in a Langevin clock, or related to some quantum process (e.g. flavour oscillations of neutrinos; Ahluwalia and Burgard 1996, Knox 2010), or even related to the periods of binary systems of black holes, are a necessary component of measurements of proper time (or, more generally, parameters along trajectories that are frame-independent). But oscillations require a time parameter, and as we can see from equation (1), the supertranslation symmetry of superspace means that the previously temporal dimension picks up a fermionic component. And these extra dimensions, which, by the supertranslation transformation, are equivalent to the ordinary Minkowski dimensions, have no finite extension.

Whether or not this means that the super-surveyor hypothesis is doomed in superspace is unclear, although the absence of a sensible notion of an oscillation means, at the very least, we will need to come up with a radically new method of surveying the supermetric if the hypothesis is to be satisfied. A more detailed analysis is beyond the scope of this chapter; the aim here was to assess the verdict that Earman, Brown, and Knox would pass on the spatiotemporality of superspace. What is clear is that we cannot fall back on phenomenological principles in superspace in the way that we could in special relativity, in order to bootstrap our way up to a theory that satisfies the (super-)surveyor hypothesis. The relativity principle on its own does not determine the invariant quantity. And the efficacy of any sort of surveyor hypothesis depends crucially on this. Simply knowing about the structure of, say, the super-Minkowski metric does not immediately give us generic (i.e. dynamics-independent) chronometric facts about the behaviour of superfields—in superspace, as in certain general relativistic solutions, we do indeed need to get our hands dirty with the physical details of individual superfields and, in particular, with the details of the theory that point towards what, if any, the physically relevant invariant quantity is.

The upshot of this uncertainty about whether superspace is an operational spacetime is, however, that prescriptions due to Earman, Brown, and Knox do not guarantee that superspace will count as an operational spacetime, even though they all unequivocally categorize superspace as a theoretical spacetime.

5.5 Conclusion

This chapter had two primary aims: (i) to argue that superspace (equipped with suitable geometry) is the correct theoretical spacetime setting for supersymmetric field theories and (ii) to demonstrate that merely identifying theoretical spacetime structure does not guarantee operational information about how measuring devices (and therefore matter fields in general) would behave, because of the possibility of failure of the super-surveyor hypothesis. As a result of this severance, a large class of spatiotemporally oriented metaphysical claims will have to be reassessed.

A broader secondary aim was to present a first attempt at identifying some of what might go into an interpretation of a spacetime compatible with (or derived from, as the case may be) a theory of quantum gravity. It is often (I think sensibly) argued that our final theory of quantum gravity, whatever it may be, will make it necessary to reject much of what we take for granted in such debates (geometric features of the space of independent coordinates, for example). But absent such a theory, our speculation and philosophical insights are best grounded in extant debates and points of view on the philosophy of spacetime physics. As Pooley (forthcoming, 183) maintains, "one is likely to do more justice, not less, to the conceptual novelty of [general relativity] by seeking as much common ground with previous theories as possible". Even so, it is clear that our most basic version of a supersymmetric theory is highly counter-intuitive, involving references to superfields, Grassmann algebra-valued coordinates and measures, and integrating super-Lagrangians. It might just be the case, however, that desperate spacetimes call for desperate measures.

References

Ahluwalia, D. V. and C. Burgard (1996). Gravitationally induced neutrino-oscillation phases. *General Relativity and Gravitation 28*(10), 1161–1170.

Anandan, J. (1980). On the hypotheses underlying physical geometry. *Foundations of Physics 10*(7–8), 601–629.

Anderson, J. L. (1967). *Principles of Relativity Physics*. New York: Academic Press.

Arntzenius, F. (2012). *Space, Time and Stuff*. Oxford: Oxford University Press.

Barrett, J. A. (1999). *The Quantum Mechanics of Minds and Worlds*. Oxford: Oxford University Press.

Becker, K., M. Becker, and J. H. Schwarz (2006). *String Theory and M-theory: A Modern Introduction*. Cambridge: Cambridge University Press.

Belot, G. (2000). Geometry and motion. *British Journal for the Philosophy of Science 51*(4), 561–595.

Brown, H. R. (2005). *Physical Relativity*. Oxford: Oxford University Press.

Brown, H. R. and J. Read (forthcoming). The dynamical approach to spacetime. In E. Knox and A. Wilson (Eds.), *The Routledge Companion to Philosophy of Physics*. London: Routledge.

Buchbinder, I. L. and S. M. Kuzenko (1998). *Ideas and Methods of Supersymmetry and Supergravity*. Bristol: Institute of Physics Publishing.

Callender, C. (2017). *What Makes Time Special?* Oxford: Oxford University Press.

DeWitt, B. (1992). *Supermanifolds*. Cambridge: Cambridge University Press.

Earman, J. (1989). *World Enough and Space-Time: Absolute versus Relational Theories of Space and Time*. Cambridge, MA: MIT Press.

Ehlers, J., F. A. Pirani, and A. Schild (2012). Republication of: The geometry of free fall and light propagation. *General Relativity and Gravitation 44*(6), 1587–1609.

Fletcher, S. C. (2013). Light clocks and the clock hypothesis. *Foundations of Physics 43*(11), 1369–1383.

Friedman, M. (1983). *Foundations of Space-Time Theories: Relativistic Physics and Philosophy of Science*. Princeton, NJ: Princeton University Press.

Friedman, M. (2001). *Dynamics of Reason*. Stanford, CA: CSLI Publications.

Huggett, N. and C. Wüthrich (2013). Emergent spacetime and empirical (in)coherence. *Studies in History and Philosophy of Modern Physics 44*(3), 276–285.

Isenberg, J. and J. Nester (1980). Canonical gravity. In A. Held (Ed.), *General Relativity and Gravitation: Vol. 1. One Hundred Years after the Birth of Albert Einstein*, pp. 23–97. New York: Plenum Press.

Knox, E. (2010). Flavour-oscillation clocks and the geometricity of general relativity. *British Journal for the Philosophy of Science 61*(2), 433–452.

Knox, E. (2013). Effective spacetime geometry. *Studies in History and Philosophy of Modern Physics 44*(3), 346–356.

Knox, E. (2019). Physical relativity from a functionalist perspective. *Studies in History and Philosophy of Modern Physics 67*, 118–124.

Maudlin, T. (2012). *Philosophy of Physics: Space and Time*. Princeton, NJ: Princeton University Press.

Menon, T., N. Linnemann, and J. Read (forthcoming). Clocks and chronogeometry: Rotating spacetimes and the relativistic null hypothesis. *British Journal for the Philosophy of Science*. https://doi.org/10.1093/bjps/axy055.

Minkowski, H. (1909). Space and Time. *Annual Report of the German Mathematician Association 18*, 75–88.

Misner, C. W., K. S. Thorne, and J. A. Wheeler (1973). *Gravitation*. London: Macmillan.

Pooley, O. (2017). Background independence, diffeomorphism invariance, and the meaning of coordinates. In D. Lehmkuhl, G. Schiemann, and E.Scholz (Eds.), *Towards a Theory of Spacetime Theories*, pp. 105–143. Basel: Birkhäuser.

Pooley, O. (forthcoming). *The Reality of Spacetime*. Oxford: Oxford University Press. Work in progress. Sent in personal communication.

Read, J., H. R. Brown, and D. Lehmkuhl (2018). Two miracles of general relativity. *Studies in History and Philosophy of Modern Physics 64*, 14–25.

Read, J. and T. Menon (2019). The limitations of inertial frame spacetime functionalism. *Synthese*. https://doi.org/10.1007/s11229-019-02299-2.

Rogers, A. (2007). *Supermanifolds: Theory and Applications*. Hong Kong: World Scientific.

Saunders, S. (2003). Physics and Leibniz's principles. In K. Brading and E. Castellani (Eds.), *Symmetries in Physics: Philosophical Reflections*, pp. 289–307. Cambridge: Cambridge University Press.

Skow, B. (2007). What makes time different from space? *Noûs 41*(2), 227–252.

Suppes, P. (1961). A comparison of the meaning and uses of models in mathematics and the empirical sciences. In H. Freudenthal (Ed.), *The Concept and the Role of the Model in Mathematics and Natural and Social Sciences*, pp. 163–177. New York: Springer.

Wallace, D. (2019). Who's afraid of coordinate systems? An essay on the representation of spacetime structure. *Studies in History and Philosophy of Modern Physics 67*, 125–136.

Weinberg, S. (2000). *The Quantum Theory of Fields vol. III: Supersymmetry*. Cambridge: Cambridge University Press.

Wess, J. and J. Bagger (1992). *Supersymmetry and Supergravity*. Princeton, NJ: Princeton University Press.

Zwiebach, B. (2004). *A First Course in String Theory*. Cambridge: Cambridge University Press.

6
Thinking about Spacetime

David Yates

Abstract

Several different quantum gravity research programmes suggest, for various reasons, that spacetime is not part of the fundamental ontology of physics. This gives rise to the problem of empirical coherence, which I frame in terms of entailment: how could a non-spatiotemporal fundamental theory entail spatiotemporal evidence propositions? Solutions to this puzzle can be classified as realist or antirealist, depending on whether or not they posit a non-fundamental spacetime structure grounded in or caused by the fundamental structure. These approaches place different constraints on our everyday concepts of space and time. Applying lessons from the philosophy of mind, I argue that only realism is both conceptually plausible and suitable for addressing the problem at hand. I suggest a role-functionalist version of realism, which is consistent with both grounding and causation, and according to which our everyday concepts reveal something of the true nature of emergent spacetime.

6.1 Introduction

Several different quantum gravity research programmes suggest that spacetime is not part of the fundamental ontology of physics. The reasons vary from case to case, as does the extent to which they suggest a non-spatiotemporal fundamental ontology.[1] But the central idea, and all I need to set up the arguments of this chapter, is that the fundamental entities posited by the theories in question seem to inhabit spaces whose structures are radically different from the familiar spacetime of general relativity (GR). In loop quantum gravity (LQG), for instance, which is obtained by applying canonical quantization procedures to the Hamiltonian formulation of GR, it's very hard to identify spatiotemporal properties like locality. In LQG, the fundamental structures are spin networks, which are represented

[1] See Huggett and Wüthrich (2013) for details of these cases, graded according to how radically non-spatiotemporal their fundamental ontologies seem to be.

David Yates, *Thinking about Spacetime* In: *Philosophy Beyond Spacetime: Implications from Quantum Gravity.*
Edited by: Christian Wüthrich, Baptiste Le Bihan, and Nick Huggett, Oxford University Press.
DOI: 10.1093/oso/9780198844143.003.0006

as graph structures in which the nodes and edges are associated with quantum spin states. These spin states are eigenstates of operators that can be interpreted as area and volume operators, and which have discrete spectra, so the fundamental LQG space is not the smooth differentiable manifold of GR. Localization in LQG spacetime must be understood in terms of interaction between other quantum fields and the gravitational field as described by LQG, rather than in terms of position in a background spacetime, as in standard quantum field theory.[2] These properties of LQG—the discrete nature of LQG spacetime and the interactive theory of localization it requires—are certainly radical departures from a continuous background spacetime, but they don't suggest the disappearance of spacetime in LQG. They suggest that spacetime is fundamentally discrete and dynamic.[3] However, other aspects of LQG are more problematic.

The nodes of spin networks can be interpreted as quanta of volume and the edges as quanta of area, with adjacency relations definable within a given spin network. However, adjacent nodes in a spin network map onto areas that can be arbitrarily far apart in general relativistic spacetime. Hence, there is no relation corresponding to the classical locality relation in the fundamental LQG structure, and no clear way of identifying the properties and relations of LQG with ordinary spatiotemporal properties and relations. Furthermore, when we factor in that the fundamental space ought to be in a *superposition* of spin network states, rather than an eigenstate thereof, the problem becomes more pressing still, as the terms of the superpositions have different structures, so what is local in one need not be local in another. These problems are jointly known as the problem *of disordered locality*.[4] As has been noted, disordered locality in LQG seems to preclude straightforward spatiotemporal interpretations of LQG operators, according to which the nodes and edges in spin networks are quanta of *spatial* volume and area respectively, and adjacency in a spin network is *spatial* contiguity.[5]

Similar issues arise in string theory, in virtue of T-duality. T-dual theories are defined on *compactified* spaces, in which at least one dimension has the topology of a circle, but where the radius assigned by one of the pair of theories is the reciprocal of that assigned by the other.[6] T-duality is a mapping that preserves all observables, but such that each of the pair of theories attributes a different radius to a compactified dimension of the space—'target space'—in which the string-theoretic dynamics play out. Philosophers of physics typically embrace the *common core* approach, according to which the physical content of T-duals should

[2] Rovelli and Vidotto (2014). [3] See for instance Huggett and Wüthrich (2013, 277–278).
[4] Following Markopoulou and Smolin (2007).
[5] Huggett and Wüthrich (2013); Wüthrich (2017); Lam and Wüthrich (2018).
[6] See Huggett and Wüthrich (2013, 280–281) for an informal account.

be seen as that which is common to both theories.[7] The physical geometry of target space should thus be regarded as indeterminate. Nick Huggett argues on this basis that target space can't be identified with *phenomenal* space, the space we take ourselves to occupy and about which GR is our best theory.[8] GR spacetime has a *determinate* geometry, and so can't be identified with target space, because nothing can have both a determinate and an indeterminate geometry. As with LQG, there's good reason to think that the fundamental ontology of string theory is non-spatiotemporal, in the sense that its dynamics play out in a structure that can't be identified with GR spacetime.

6.1.1 The Problem of Empirical Coherence

The apparent disappearance of spacetime in quantum theories of gravity gives rise to a problem of *empirical coherence*. To a first approximation, the problem can be stated as a question: if the elements of the fundamental ontology don't occupy spacetime, then how can fundamental theories be justified by observation of things that do, such as meters, pointers, and dials—the so-called local beables of the ontology?[9] This problem has already been much discussed in the context of wavefunction realism, and it will be useful to frame the present debate in the terms already established there. Briefly, there are good grounds for treating the wavefunction as a physical field, but it can't be a field on 3D space, because there aren't enough degrees of freedom there to distinguish between distinct quantum states.[10] Rather, the wavefunction must inhabit configuration space, which has at least 3N dimensions, where N is the number of particles in the universe. None of these 3N dimensions corresponds to any of the three dimensions of space we intuitively take our world to have, so if configuration space is fundamental, how can there be local beables? I shall frame the problem of empirical coherence as follows:[11]

1. A fundamental theory is empirically coherent only if it entails evidence propositions.
2. Evidence propositions attribute spatiotemporal properties to local beables.
3. The fundamental ontology of physics is non-spatiotemporal.

[7] See for instance Le Bihan and Read (2018); Matsubara and Johansson (2018); De Haro (2019).
[8] Huggett (2017).
[9] Healey (2002); Maudlin (2007); Huggett and Wüthrich (2013).
[10] See the appendix to Ney (2012) for details.
[11] Following Maudlin (2007); see also Ney (2015).

I assume for the purposes of this chapter that (3) is true.[12] Premise (1) is uncontroversial. An evidence proposition as understood here is just a proposition describing a potentially observable state of a physical system, such as 'the pointer moved to position 5 on the dial'. A theory that failed to entail evidence propositions could not be confirmed by observation. Premise (2) is intuitively plausible: evidence propositions *seem* to be claims involving spatiotemporal configurations of observables. But by (3), physics doesn't have local beables in its fundamental ontology, so how could it entail evidence propositions? There's no problem if the fundamental ontology is spatiotemporal, for in that case it contains fundamental local beables, whose dynamics determine the behaviour of the non-fundamental things they compose. By contrast, non-spatiotemporal fundamental theories seem to preclude their own confirmation, and this is the sense in which they are threatened by empirical incoherence.

6.1.2 Realist and Antirealist Solutions

As I have set things up, the problem of empirical coherence is one of entailment: how could a non-spatiotemporal fundamental theory entail evidence propositions attributing spatiotemporal properties to local beables? In what follows, I will classify responses to this problem as *realist* or *antirealist*, depending on whether or not they are committed to a non-fundamental spacetime in addition to the ontology of fundamental physics. My usage of 'realism' and 'antirealism' here departs from traditional usage in the philosophy of science. Some positions classified as antirealist in my usage would be classified as realist according to certain traditional measures, for instance whether or not there are mind-independent spatiotemporal truths. My distinction concerns only the existence or otherwise of a non-fundamental spacetime as truthmaker for such propositions, and should be taken to have no further ontological implications. For brevity, I will henceforth use '*S*-concepts' to refer to spatiotemporal concepts, and '*S*-propositions' for propositions featuring such concepts.[13]

In order for a non-spatiotemporal fundamental theory to have observational consequences, we will need some way of connecting the language of that theory with the *S*-concepts that we use to describe possible observations. Clearly, it's impossible to know what form this connection will take without a final theory. My aim in this chapter is to explore the space of possibilities by focusing on the

[12] Some solutions to the problem of empirical coherence that purport to show how empirical confirmation might work given non-spatiotemporal fundamental ontologies in my view tacitly presuppose that (3) is false, and so fail to deliver on their promises. I return to this in §6.2.3.

[13] I shall use UNDERSCORE to refer to specific concepts.

spatiotemporal side of the relation. According to realism, *S*-propositions describe non-fundamental states in the *derivative* ontology of the fundamental theory, with ordinary objects and their spatiotemporal properties *emergent* in relation to the fundamental.[14] Perhaps the fundamental structure causes a non-fundamental spacetime structure to emerge in the appropriate classical limit, or perhaps it synchronically grounds that structure. For realists, the challenge is to give an account of the emergence relation itself, which will then bridge the fundamental and the spatiotemporal, and render (1)–(3) consistent. Antirealism eschews robust emergence in favour of a deflationary approach, according to which *S*-concepts don't attribute the kind of properties we ordinarily take them to. As we shall see, there are several distinct forms of antirealism, but they all have it in common that they deny premise (2). The challenge for antirealists is to give an account of *S*-concepts that is both conceptually plausible *and* which offers the promise of a solution to the problem of empirical coherence. My primary aim in this chapter is to argue that these aims cannot jointly be met, thereby motivating realism.

6.2 Spacetime Antirealism

A preliminary defence of spacetime antirealism is to be found in Ney's work on wavefunction realism. Given wavefunction realism, if 3D space is real, then it must somehow be derived from 3*N*-dimensional configuration space. Ney suggests that we can do without derived space:

> I am not arguing that on a straightforward, ontological reading of realist versions of quantum mechanics that there are not rocks . . . However, I claim that this does not imply that there is anything that is genuinely rock-shaped (if this implies occupying locations in a physical, three-dimensional space) or that moves in precise trajectories through such a space. Since we can allow that the wavefunction simulates the behaviour of something in a three-dimensional space, we have a way of making sense of our observations in terms of the language of the theory.[15]

[14] See Ney (2012) for full discussion of this position in the context of the wavefunction realism debate. In my current usage, emergence is consistent with physical realization. If general relativistic spacetime were grounded in a non-spatiotemporal fundamental ontology in a fully transparent way, then it would count as emergent in this usage. This usage of the term 'emergence' is common among philosophers of science, but less so among metaphysicians, who tend to think of the sort of emergence that gives rise to genuinely novel dependent entities as requiring in principle non-deducibility or brute supervenience. See Yates (2016) for more on this issue.

[15] Ney (2012, 550). Note that in her (2017), Ney defends the possibility of *deriving* ordinary spacetime from configuration space mereologically, so isn't committed to antirealist solutions.

Ney's suggestion is that we can identify physical objects with states of the wavefunction on configuration space, provided we give up on the idea that they literally have spatiotemporal properties like location, shape, and so forth. Thus, Ney is an antirealist in the current sense, seeking to explain empirical confirmation without having to derive a non-fundamental spacetime from fundamental physics. According to Ney, confirmation requires only:

> [A] difference that can be grounded in the ontology of the quantum theory between what we refer to as 'the pointer pointing to up' and 'the pointer pointing to down'. But the wave function realist is free to trace the difference in what we refer to as 'the pointer pointing to up' and 'the pointer pointing to down' to differences in the state of the wave function in the two scenarios.[16]

On this view, confirmation doesn't require observation of distinct states of local beables, but simply observation of distinct states, which may then be understood as being states of the wavefunction on configuration space. What we ordinarily describe as someone perceiving successive states of a measuring instrument in 3D space is really the wavefunction as a whole evolving from one state to another. Both perceiver and instrument are identified with patterns in the wavefunction, and despite appearances to the contrary, neither is located in 3D space. In order for this kind of confirmation theory to work, we need a theory of *S*-concepts according to which *S*-propositions are in principle deducible from the fundamental theory, but are not attributions of spatiotemporal properties to local beables. There are several options available, all of which involve treating *S*-concepts as *non-transparent*.

6.2.1 Transparent vs. Non-Transparent Concepts

I will say that a concept is *transparent* if deploying that concept involves thinking about its referent *as* the bearer of (at least some of) its essential properties. Transparent concepts, if any such there be, give us a kind of *a priori* access to the natures of their referents, in the sense that thinking about an entity under a transparent concept is thinking about it as it is essentially. Non-transparent concepts pick out their referents in terms of inessential properties.[17] Our everyday concept water is *non*-transparent: to think ordinary water-thoughts is to think about a natural kind in terms of its surface features, such as being colourless, odourless, and tasteless. The chemical essence of water is only revealed by scientific

[16] Ney (2015, 3210).

[17] Obviously, on such an account, there can be no transparent concepts of things that don't have essential natures. Goff (2017, 97–100) is particularly clear on transparent concepts.

investigation, and is not accessible to us *a priori*. Many suppose that in contrast to natural kind concepts, *phenomenal* concepts are transparent. If so, then to think about pain (say) in terms of the way it feels is to represent a specific mental state as it is essentially. The essence of pain is revealed by experiencing what it's like, and this experience enables us to form a phenomenal concept PAIN, which then refers to the property of being in pain by means of its essential phenomenology.[18]

Now *S*-concepts are intuitively transparent: something of the nature of sphericality is revealed to us, we might suppose, in deploying the concept SPHERICAL. Chalmers offers several independent twin-Earth-style arguments against the transparency of *S*-concepts.[19] Chalmers suggests[20] that *S*-concepts might pick out "classical properties in a Newtonian world, relativistic properties in a relativistic world, and string-theoretic properties in a string-theory world". For Chalmers, *S*-concepts are functional, and pick out elements of the fundamental ontology as the occupants of contingent roles, and not as the properties they are in themselves. This sort of position is known, variously, as *realizer*, *conceptual*, or *filler* functionalism; I will use the term 'realizer functionalism'. In the remainder of this section I will consider two versions of realizer functionalism about *S*-concepts, both of which are attributable to Chalmers: *phenomenal* and *theoretical*, depending on the kind of role specification involved.

6.2.2 Phenomenal Realizer Functionalism

Most would accept that if we are in the Matrix, then almost all of our beliefs are false—it's not true that I'm sitting here typing this sentence if I'm attached to a computer that's feeding me all the sensory inputs that cause me to have experiences as of sitting here typing it. Spacetime realizer functionalism, for Chalmers, forms part of a challenge to this orthodoxy. In *phenomenal* spacetime realizer functionalism, SPHERICALITY is the concept of whatever normally causes sphere-ish experiences. If phenomenal realizer functionalism is true, then if I am in the Matrix and have an experience as of a spherical ball, I can truly say that I see something spherical, provided whatever part of the Matrix program it is that normally causes that kind of phenomenology is the cause of my experience. The bearers of spatial properties are genuine physical objects, albeit objects whose

[18] I need not commit to the transparency of phenomenal concepts for present purposes, as my aim is merely to give the reader a sense of what's involved in affirming or denying conceptual transparency in general. See Lewis (1995) for compelling arguments that materialists ought to reject the transparency of phenomenal concepts.

[19] Chalmers (2012, 325–336); Chalmers (2019). I lack the space to engage with Chalmers' twin-Earth arguments here, but the position suggested in §6.3.1 predicts twin-Earth cases for partially transparent *S*-concepts.

[20] Chalmers (2012, 325–326).

fundamental proper parts are bits of code in the Matrix mainframe. The things we think of as fundamental—electrons, photons, and quarks, for instance—are themselves composed of bits, and compose into further digital objects. Some of these latter have the complex property that normally causes sphere-ish experiences in Matrix dwellers, and are therefore spherical objects.

An immediate worry is that things that aren't spatiotemporally located can't cause our experiences, rendering phenomenal realizer functionalism of no use in addressing the problem of empirical coherence. I won't address this issue here, and assume that on a broad enough notion of causality, it makes sense to suppose that our experiences as of things located in spacetime can be caused by things that are not themselves spatiotemporally located. Can *S*-concepts be functionalized in the way this version of phenomenal realizer functionalism requires? For my part, I don't think they can. Readers familiar with the history of the philosophy of mind will be aware of the reasons for the decline of logical behaviourism. According to a simple version of that theory, individual mental states have *a priori* dispositional analyses—the belief that it's raining is the disposition to take your umbrella if you go outside, to say 'yes' when asked if it's raining, and so on. The problem with such analyses stems from the holism of the mental: even if you believe it's raining, you're only disposed to take your umbrella with you if you go outside *if you also want to stay dry*. But wanting to stay dry, *qua* mental state, will then be analysed in terms of the disposition to take your umbrella… if you also think it's raining. Which dispositions you have depends on the totality of your mental states, so no individual mental state suffices *a priori* for any specific behavioural disposition. A similar problem arises on the simple version of phenomenal realizer functionalism sketched above.

What phenomenology is associated with a given *S*-concept? What does being in contact look like, or in relative motion? Focus on the latter. The phenomenology associated with being in motion relative to a perceiver depends on many things, including: (i) what shape the object is, (ii) where the light sources are located in relation to object and perceiver, (iii) the direction of motion relative to the perceiver, (iv) the current distance between perceiver and object, and (v) the spatial relations between the object and other things in the perceiver's environment. But (i)–(v) all involve further *S*-concepts. Another example: what do two cylindrical rods look like, placed a metre apart endwise? It depends where the perceiver is located in relation to the rods, the relative size and motion of the rods and perceiver, and so forth. In short, the way things that fall under a given *S*-concept appear to a given perceiver depends on the other *S*-concepts that object and perceiver fall under. A given *S*-concept *C* can't refer to whatever normally causes *C*-ish phenomenology, because there's no way to specify *C*-ish phenomenology independently of the other *S*-concepts that the perceiver and the object perceived fall under. Just as behavioural dispositions can't be paired up 1-1 with mental concepts, so the phenomenology associated with falling under S-concepts is associated not with individual concepts, but with complex conjunctions thereof.

This leaves room for an holistic version of phenomenal realizer functionalism, according to which S-concepts are jointly defined in terms of the phenomenology associated with falling under many S-concepts at once. On this theory, we refer to complex fundamental states by means of complex S-propositions, with the level of complexity dependent on how much we have to build into the proposition to describe a situation with a unique spacetime phenomenology for a perceiver embedded within it in a particular way. Chalmers offers remarks that are consistent with holistic phenomenal realizer functionalism:

> [S]patial concepts pick out that manifold of properties that serves as the normal causal basis of a corresponding manifold of properties in our spatial experience. Temporal concepts pick out that manifold of properties that serves as the normal causal basis of a corresponding manifold of properties in our temporal experience.[21]

Gather the manifold of properties (intrinsic and relational) that jointly cause complex spacetime phenomenology together into a set M. Schematically, to think of something as spherical is then to think of it as having a certain property $P \in M$ such that P coinstantiated with the properties in $X \subseteq M$ normally causes sphere$_X$-ish phenomenology; and so on for the other subsets of M. This is quite some level of conceptual sophistication to attribute to the folk. Deploying a concept like SPHERICALITY involves thinking of an object in terms of how it would appear if it fell under that concept together with certain other concepts of the same kind—enough to correspond to a determinate phenomenology and secure unique reference to the relevant cause property. This isn't a knock-down objection, but it does make holistic phenomenal functionalism look somewhat implausible. These are *our* S-concepts, and while on realizer functionalism we don't have *a priori* access to the natures of their referents, we should surely have (at least some) *a priori* access to the manner in which they refer. So, when you think of something as spherical, are you thinking about it in terms of the various ways it would look if you and its bearer also fell under various combinations of further S-concepts? That doesn't seem to me to be what I am doing. Phenomenal realizer functionalism in any form also ties cognitive contents to specific sensory modalities, and so entails (for instance) that blind or partially sighted people can't think the same spatiotemporal thoughts as normally sighted people, because their

[21] Chalmers (2012, 335).

concepts involve different phenomenological reference-fixers.[22] A more plausible theory, to my mind, is that when we attribute sphericality to something, we are thinking about what the results would be if we measured the distance from its outer boundary to its centre—but not, of course, about the phenomenology associated with those measurement outcomes.

A further problem for phenomenal functionalism, considered as a solution to the problem of empirical coherence, is that it requires bridge principles connecting the non-spatiotemporal fundamental theory with the relevant phenomenology, but it's very difficult to see how such principles could be supplied without already having solved the problem at hand. In order to avoid a possible misunderstanding, first consider a *prima facie* similar problem that arises in spatiotemporal theories, which of course don't suffer from the problem of empirical coherence. In cases where we are able to deduce evidence propositions from a theory, we still need ways of determining, on the basis of our conscious experiences, whether those propositions are true. This conceptual competence need not take the form of phenomenal analyses of the relevant concepts, but grant for the sake of argument that it does. The connection between theory and phenomenology can then be established in two steps: (i) deduction of evidence propositions from the theory, and (ii) functional analysis of the concepts featured in those propositions in phenomenal terms. The problem for *non*-spatiotemporal theories, however, runs deeper.

If phenomenal analysis is to render *S*-propositions deducible from a non-spatiotemporal theory, we need a way of correlating arbitrary states of that theory's ontology with the phenomenology that those states typically bring about. It's no use analysing SPHERICALITY as the concept of whatever normally causes sphere-ish experiences if we don't know which state of the fundamental ontology that is. The problem now is that this correlation must be established *prior* to the derivation of any *S*-propositions from the theory: our apparent inability to deduce such propositions in quantum theories of gravity is the problem that phenomenal functionalism is supposed to solve. However, everything we think we know about perception relies on our being agents with sensory receptors located in a spatiotemporal world. Our perceptual systems, we think, work by means of sensory receptors that convert physical inputs—for instance, photons or air pressure waves—into electrical signals that our brains use to construct representations of their causes. Following Chalmers, suppose that a neural correlate of consciousness is

[22] In terms of 2D semantics, the spatiotemporal thoughts of blind folk at our world will have the same secondary intensions as, but different primary intensions from, the thoughts of sighted folk. I assume here that for cognitive purposes, the primary intensions are what matter.

"[a] a neural representational system *N* such that representation of a content in *N* directly correlates with representation of that content in consciousness".[23] Suppose now that we have made sufficient progress in neuroscience to identify the neural correlates of arbitrary phenomenal states. To predict how a *non*-spatiotemporal fundamental state will appear to creatures like us, we would *already* need a way of deducing *S*-propositions from the fundamental theory, because the neural correlates of phenomenal states are spatiotemporal.

Phenomenal realizer functionalism does have the resources—setting aside the question of its conceptual plausibility—to explain how *S*-concepts could refer to the elements of a non-spatiotemporal fundamental ontology. However, if we have no way of finding out *which* fundamental properties they refer to, we are no closer to explaining how *S*-propositions could be deducible from the fundamental theory. Phenomenal realizer functionalists need to explain how, in principle, we could identify which fundamental properties bring about the various types of spacetime phenomenology that define our *S*-concepts, but without relying on spatiotemporal theories of perception. In response to this, it might be suggested that holistic phenomenal realizer functionalism could achieve this correlation by mapping the structure of our spacetime phenomenology directly onto the structure of the fundamental theory. In effect, this would treat *S*-concepts as defined by means of a phenomenological *theory*, and explain the deducibility of *S*-propositions from the fundamental theory in terms of inter-theoretic reduction. I shall now discuss what I take to be a far stronger version of theoretical realizer functionalism, based not on phenomenology but on GR, and argue that it fails as a solution to the problem of empirical coherence for reasons that would also apply to the phenomenal theory just sketched.

6.2.3 Theoretical Realizer Functionalism

Chalmers hints at a theoretical version of realizer functionalism about *S*-concepts, but doesn't really flesh it out.[24] In this section I suggest a way to do so. Einstein's field equation, in its standard formulation, is a set of tensor equations that relate the matter distribution to the local spacetime curvature. The *metric tensor* describes all the geometric properties of spacetime, and can be used to define properties such as volume, distance, and intrinsic curvature. The field equation of

[23] Chalmers (2000, 20). Direct correlation is understood in terms of minimal sufficiency: representation of the given content in *N* is sufficient for its representation in consciousness, and no proper part of *N* is sufficient. I lack the space to go into detail, and appeal to this notion here only for illustrative purposes.

[24] See for instance Chalmers (2012, 332–334, 418–419). A version of theoretical spacetime functionalism is also developed in Lam and Wüthrich (2018); I return to their views presently.

GR relates the metric to the distribution of mass-energy by means of the *stress-energy tensor*, and expresses a mutual lawlike dependence between the matter distribution throughout spacetime and the metric structure of spacetime itself. Whether there is a causal relationship between the matter distribution and the metric is controversial, but if there is, it's not that the matter distribution *brings it about* that spacetime has a certain metric structure, because the specification of the matter distribution in the stress-energy tensor *presupposes* a metric. A more natural interpretation is that the field equation tells us that:

> [G]iven both a certain intrinsic geometry for spacetime and a specification of the distribution of mass-energy throughout this spacetime, the joint description is the description of a general-relativistically possible world only if the two descriptions jointly obey the field equation.[25]

Eleanor Knox appeals to GR to defend functionalism about the concept SPACETIME.[26] On her view, SPACETIME is the concept of "whatever occupies the role of defining a structure of inertial frames". Inertial frames are those with respect to which bodies not acted on by a force travel at uniform velocity, and within which the laws of physics take a uniquely simple form. Knox argues that the way the metric field couples to matter fields guarantees that the local symmetries of the metric are the same as the local symmetries of the dynamical laws. This in turn explains why the metric governs the behaviour of bodies that obey those laws. Given that the metric field occupies the spacetime role, Knox's functionalism entails that it should be identified with spacetime. Now, if the fundamental ontology is non-spatiotemporal, then the metric field isn't fundamental. Knox's brand of realizer functionalism therefore identifies spacetime with a non-fundamental, derived structure, and so doesn't provide an antirealist solution to the problem of empirical coherence. To do that, we need to embrace theoretical realizer functionalism about the *S*-concepts that appear in our evidence propositions, and not just the concept SPACETIME.

According to the Ramsey-Lewis semantics, theoretical terms are defined by their places in a theoretical structure.[27] We can write a theory T schematically as follows: $T[\langle t_1, \ldots, t_n\rangle]$, where the ordered n-tuple $\langle t_1, \ldots, t_n\rangle$ denotes the theoretical (or t-) terms to be defined by means of T. The remainder of T's vocabulary (the o-language) consists of terms whose meanings are given independently of T. We get the Ramsey-sentence of T by replacing $\langle t_1, \ldots, t_n\rangle$ with bound variables: $\exists!\langle x_1, \ldots, x_n\rangle\ T[\langle x_1, \ldots, x_n\rangle]$. The meaning postulates of T are the Carnap-sentence: $\exists!\langle x_1, \ldots, x_n\rangle\ T[\langle x_1, \ldots, x_n\rangle] \rightarrow T[\langle t_1, \ldots, t_n\rangle]$, and a sentence saying that $\langle t_1, \ldots, t_n\rangle$ is empty if T is not uniquely realized. Lewis appeals

[25] Sklar (1976, 75). [26] Knox (2019). [27] See Lewis (1972) for details.

to these sentences to offer an explicit definition of the t-terms: $\langle t_1, \ldots, t_n \rangle =_{df}\ \iota\langle x_1, \ldots, x_n \rangle\ T[\langle x_1, \ldots, x_n \rangle]$. Less formally, $\langle t_1, \ldots, t_n \rangle$ refer to the unique n-tuple satisfying T, or (less formally still) to the n-tuple that occupies the T-roles. S-concepts appear as theoretical terms in GR, so why not Ramsify them relative to that theory and say that they refer to the fundamental occupants of the GR roles?

This proposal doesn't entail that S-concepts are *entirely* theoretical, defined in ways that are beyond the ken of anyone who doesn't know GR. S-concepts already appear in our folk conception of the world, and in some respects at least, GR can be seen as an extension of that conception. Concepts like DURATION, DISTANCE, CONTIGUITY, SHAPE, AREA, and VOLUME all figure in our everyday conception of space and time, by way of the range of folk-physical platitudes we are inclined to accept—for instance, the transitivity of 'later than', or the idea that only contiguous objects can interact. If there's enough in common between our folk spacetime theory and GR, then the S-concepts in question can plausibly be seen as extended by GR, not re-defined wholesale. The issue of how wrong a theory can be before reference failure occurs is of course controversial, but it's at least plausible that our folk conception of spacetime gets *enough* right for GR to count as a conservative extension thereof.[28] On the present account, then, when we deploy ordinary S-concepts to note the position of the meter on the dial, we refer to fundamental non-spatiotemporal properties, by means of the theoretical roles the corresponding scientific concepts assign to something via GR.

Theoretical realizer functionalism explains the in principle deducibility of S-propositions from a non-spatiotemporal fundamental theory, on the assumption that states expressible in the language of that theory occupy GR roles. We don't need psychophysical bridge principles in this case—an S-concept such as LOCALITY refers to the fundamental occupant of the locality role as defined in GR, so deducibility is a matter of mapping the fundamental role occupants onto the GR roles.[29] The problem for theoretical realizer functionalism, as a solution to the problem of empirical coherence, is that it ceases to work just when (and if so because) we have good reason to regard the fundamental ontology as non-spatiotemporal in the first place. If the t-terms of GR are to refer to non-spatiotemporal properties in a quantum gravity ontology, then we will need to recover enough of the nomic structure of GR from the fundamental theory that GR role specifications pick out fundamental role occupants. But the more closely the structure of GR can be recovered from a quantum theory of gravity, the less reason we will have for regarding that theory as non-spatiotemporal. The same issues arises, *mutatis mutandis*, for the phenomenal version of theoretical realizer

[28] I return to this point in §6.3.1.

[29] Note that this will require inter-theoretic bridge principles in order to connect GR roles with the nomic roles available in the fundamental theory. My point is that no problematic *psychophysical* bridge principles are needed in order to explain how S-propositions are deducible from the fundamental theory.

functionalism mooted at the end of §6.2.2. An example from the philosophy of mind will clarify this problem.

Suppose we Ramsify folk psychology, so that (for instance) PAIN refers to the brain state that's caused by tissue damage, and causes both the various forms of physical and verbal pain behaviour, and other psychological states such as anxiety. Similarly, ANXIETY will refer to the brain state that occupies the anxiety role, part of which involves being caused by the brain state that occupies the pain role. In short, the entire nexus of folk-psychological state terms refers holistically to a network of brain states that stand in the causal relations specified by the theory, both in relation to each other and to physical and verbal behaviour. In Lewisian philosophy of mind, given that pain = the occupant of causal role *R*, if the occupant of the pain role is brain state *B*, then pain is contingently *identical* to brain state *B*, because *B* contingently occupies pain's defining role.[30] That is to *reduce* pain to a state that is contingently mental, not to show how pain could refer to a *non*-mental state. Now suppose that no brain state comes close to occupying the folk-psychological pain role, or the roles of beliefs and desires. It follows that folk-psychological terms *won't* pick out brain states as the occupants of folk-psychological roles, because those roles aren't occupied at all, and Ramsified folk psychology is a set of false existential claims.[31] In that case, the brain may well be fundamentally non-mental, but for that reason, we won't be able to refer to brain states using Ramsified folk-psychological terms.

Theoretical realizer functionalism faces a dilemma. If Ramsified GR terms succeed in referring to the fundamental properties of a theory like LQG or string theory, then each GR role is (approximately) occupied by such a property. But in that case, the fundamental properties in question can be *identified* with GR spatiotemporal properties, in which case the relevant fundamental ontology can't be non-spatiotemporal. Conversely, if the ontology is non-spatiotemporal, then Ramsified GR terms won't pick out its elements. It's instructive to consider at this point a proposal due to Lam and Wüthrich on how a version of theoretical realizer functionalism helps us to recover spacetime in LQG.[32] They suggest that *weave states* of LQG spin networks do have properties that approximately occupy the roles of locality, area, and volume in GR. Ramsified GR terms can thus pick out the properties of such states as the (approximate) occupants of GR roles, with many different weave states capable of occupying the same roles. However, Lam and Wüthrich are clear that this class of models is unified in having at most

[30] Lewis (1994).

[31] Those familiar with the literature on eliminative materialism will recognize the inference from the falsity of folk psychology to reference-failure via the Ramsey-Lewis semantics; see for instance Churchland (1981).

[32] Lam and Wüthrich (2018).

only slightly disordered locality. In effect, the disordered locality problem that motivates the claim that LQG is non-spatiotemporal must already be solved in order for Ramsified GR terms to pick out fundamental LQG properties. What Lam and Wüthrich have shown, then, is that there is a class of models of LQG in which a theoretical realizer functionalist strategy could be made to work. But for precisely that reason, I submit, there's no obvious reason to regard the models in question as non-spatiotemporal.[33]

What if we abandon the idea that *S*-concepts are semantically functional and try to treat them as directly referring natural kind terms? Suppose for illustration that some version of the causal-historical theory of reference is true.[34] On that theory, we first baptise a natural kind such as water by ostension, and thereby fix the reference of water for all subsequent users of the concept, by means of causal relations to the initial baptism. If *S*-concepts refer directly, then we no longer need to recover the nomic structure of GR from the fundamental theory; deducibility of *S*-propositions will be secured when we can identify the fundamental kinds to which *S*-concepts directly refer. But how are we supposed to do *that*, if the fundamental ontology is non-spatiotemporal? Conversely, if concepts like CONTIGUITY, BEING 1M APART, SPHERICALITY, and so forth refer to kinds expressible in the language of a fundamental theory of quantum gravity, then why should we regard that theory as non-spatiotemporal? The disordered locality problem in LQG, for instance, seems to preclude identifying locality properties of spin networks with spatiotemporal locality, so if that problem is serious, we should expect there *not* to be fundamental LQG kinds corresponding to GR spatiotemporal concepts. As before, it may be plausible that there are models of the relevant theories in which we can pick out fundamental kinds by means of ostension to spatiotemporal states of affairs, but again, it won't be clear why we should think of those models as non-spatiotemporal.

I conclude that antirealist approaches to the problem of empirical coherence are untenable. Phenomenal realizer functionalism is conceptually implausible, and whereas theoretical realizer functionalism has greater plausibility, as a solution to the problem at hand it depends on a tacit rejection of the claim that the fundamental ontology is non-spatiotemporal.[35]

[33] I suspect that a similar dilemma arises for the claim that Ramsified GR terms could refer to string-theoretic properties in a space of indeterminate geometry, but I omit discussion for brevity.

[34] Kripke (1972).

[35] This tacit move could be made explicit, and theoretical realizer functionalism recast as a position according to which, despite appearances to the contrary, the fundamental ontology is spatiotemporal after all. The resulting position—suitably renamed—would then *dis*solve the problem of empirical coherence by rejecting (3) of §6.1.1.

6.3 Spacetime Realism

Despite my misgivings about antirealism as a solution to the problem of empirical coherence, I think it's plausible that *S*-concepts are semantically functional and defined in terms of GR. This idea is consistent with realism, because Ramsifying a theory *T* doesn't settle the issue of what kind of properties its *t*-terms refer to.[36] In particular, Ramsifying *T* doesn't commit us to treating its *t*-terms as referring to the fundamental *occupants* of roles specified by *T*. As Lewis himself notes,[37] we can also treat them as referring to a domain of second-order functionally individuated properties that are *realized* by the fundamental role occupants. On this theory, *t*-terms are still semantically functional, because they pick out their referents by means of their places in a structure, but they will also now be *transparent*, at least to the extent that they get the structure right. In short, the claim that the *t*-terms of a theory *T* are semantically functional leaves open whether their referents are role-occupants or role-properties,[38] and in the latter case, on the assumption that *T* is true, those concepts will be transparent, since they will pick out their referents by means of their defining roles. In the remainder of this section, I will first outline a version of role spacetime functionalism that's consistent with the partial transparency of *S*-concepts, and then consider whether a role-functional spacetime is the sort of structure that could be synchronically grounded in the non-spatiotemporal.

6.3.1 Role Spacetime Functionalism

There are many different versions of role spacetime functionalism, depending on which theory supplies the role specifications. For instance, we might say that SPHERICAL refers not to the unique property that gives rise to sphere-ish phenomenology, but to the second-order property of having *some* property that occupies that role. On this view, which we might call simple phenomenal role spacetime functionalism, spatiotemporal properties themselves are defined in terms of giving rise to spacetime phenomenology, in contrast to the corresponding

[36] Whittle (2008) argues at length that the Ramsey-Lewis method, while useful for giving functional analyses of concepts as part of a broader functionalist theory of properties, does not in itself constitute such a theory.

[37] Lewis (1994, 307–308). Realizer functionalism is a matter of theory choice, for Lewis, based on the not uncommon assumption that second-order functional properties are causally excluded by their realizers. That assumption, however, can be challenged in various ways; see Yates (2012) for full discussion.

[38] It should be noted that these options are not mutually exclusive: on a powers theory, for instance, semantically functional concepts such as CHARGE pick out fundamental properties as the occupants of the causal roles that individuate them. For more on the relational individuation of fundamental properties, see Yates (2018).

realizer functionalist theory, in which the phenomenology acts merely as a reference fixer to an underlying mind-independent reality.[39] An holistic version of phenomenal role functionalism about spacetime could also be given, in which spatiotemporal properties are defined in terms of the phenomenology they give rise to in various combinations with each other.

Functionalist theories are often divided into analytical (or common-sense) and scientific, depending on whether or not the theory that supplies the role specifications is *a priori*. This distinction is orthogonal to that between realizer and role functionalism. In Lewis' realizer functionalism, it's supposed to be *a priori* that PAIN refers to the unique occupant of a certain causal role because folk psychology is a *common-sense* theory of mind.[40] The theory in question, for Lewis, supplies reference-fixing descriptions that pick out brain states, but we can easily define a second-order version of Lewisian functionalism. According to this theory, the mental concept PAIN refers *a priori* to the property of being in some state that occupies the folk pain role. It follows that our ordinary mental concepts are fully transparent to their referents, because when we deploy those concepts, we thereby think of mental properties in terms of their essential natures, without remainder. Now consider *psychofunctionalism*,[41] according to which mental properties are defined by their places in an *a posteriori* scientific psychological structure. In psychofunctionalism, it is often said that the role of folk concepts is not to capture the true natures of mental properties, but to fix the reference of mental concepts, which refer to properties whose true natures are given by deep scientific psychology. But now suppose that scientific psychology is a (relatively) *conservative extension* of folk psychology—suppose, that is, that the folk know something about the essential natures of the mental properties we attribute. On this view, not all of folk psychology will be abandoned as cognitive science progresses; some folk claims will be preserved (in some form) as components of mature scientific psychology. It follows that folk-psychological concepts are *partially* transparent: they pick out mental states by specifying parts of their individuative roles. I shall now suggest that the same could be true of *S*-concepts.

In the Minkowski spacetime of special relativity, properties like shape, size, and duration vary depending on the relative motion of the observer and the objects or events in question; and in the pseudo-Riemannian spacetime of GR, the intrinsic curvature of spacetime depends on the matter distribution. However, such radical departures from common sense don't rule out that a theory is conservative, in the

[39] Schwitzgebel (2019) argues for a Kantian phenomenological conception of Matrix spacetime based on simple phenomenal role functionalism about spacetime.

[40] The role specifications are *a priori* in the limited sense that no scientific investigation is required in order to know them; possession of folk concepts is sufficient. Nothing turns on epistemic priority for my purposes.

[41] See Block (1978).

sense outlined above. As is familiar, special relativistic effects such as Lorentz contraction and time dilation are only significant between inertial frames at very high relative velocities. For observers that are not too far from being in the same inertial frame, shapes, sizes, and durations are (approximately) the same, and the same events are (approximately) simultaneous. For observers at relative velocities far from the speed of light, our common-sense conception of space and time, as reflected in our pre-theoretic *S*-conceptual structure, is approximately *true*. GR is more revolutionary than SR, as it treats spacetime as a dynamic structure rather than a background container for the dynamics. However, the point still stands: locally and for weak gravitational fields, GR spacetime approximates Minkowski spacetime, which in turn departs from common sense only at high relative velocities.

Relativity doesn't so much refute common sense as reveal its *limits*. Because common sense posits like absolute spatial and temporal intervals, and flat Euclidean geometry, are recoverable in the appropriate limits of SR and GR, it's plausible that at least some of our common-sense theory of space and time is preserved therein. Now suppose that ordinary *S*-concepts refer to role-functional properties, individuated by their places in a structure described by GR.[42] Partial transparency follows. Our ordinary ways of thinking about space and time are a mixture of semi-formal and causal principles, for instance the transitivity of 'longer than' and 'later than'; the idea of an absolute present moment; the claim that *A* and *B* are equidistant from *C* iff it takes the same time to travel between *A* and *C* as between *B* and *C* at constant velocity; spatiotemporal contiguity as a necessary condition on interaction; the principle that causes must precede their effects; and so on. If some of this structure is preserved in GR, then our ordinary ways of thinking about spacetime represent their referents as relata in (a proper substructure of) the general relativistic structure that individuates them.[43]

As long as the folk aren't completely wrong about spacetime, there's room for a functionalist theory that combines the partial transparency of *S*-concepts with the idea that spatiotemporal properties are scientific kinds. This entails the possibility of twin-Earth cases involving *S*-concepts, for we can always imagine worlds with properties that instantiate the folk structure, but at which the scientific extension is different, or even absent. Imagine a twin Earth, for instance, at which spacetime is a classical background container for the dynamics with zero curvature every-

[42] The Minkowski metric is an approximate solution to the field equations of GR in the limit of zero gravitational field, i.e. empty space, so SR is typically seen as reducible to GR. For that reason I refer only to GR as the theory that specifies the defining roles for a role-functionalist conception of spacetime. See Lam and Wüthrich (2018) for a functionalist analysis of (*inter alia*) spatiotemporal locality based on its role in GR.

[43] Note that this does not amount to what Chalmers (2012, 325) calls spatiotemporal *primitivism*, the claim that we have direct unmediated grasp of spatiotemporal properties. Our grasp of such properties, on the position outlined here, is mediated by our grasp of the structure that individuates them.

where. It's intuitively plausible that twin-Earthlings would have the same everyday *S*-conceptual framework as Earthlings, and also that the twin-Earth concept *T*-SEPARATION, for instance, refers to a different property than our SEPARATION. Our concept picks out a property of a dynamic structure whose geometry depends on the matter distribution, whereas theirs does not. However, given the possibility of a version of role functionalism in which at least some of the folk roles are incorporated into scientific theory, this doesn't suffice to show that our *S*-concepts are *fully* non-transparent. The fact that we can construct twin cases for a family of concepts suggests that there's more to their referents than meets the eye, but it doesn't follow that what meets the eye is merely superficial.[44]

6.3.2 Grounded or Caused?

In this section I will first consider the conditions under which it is possible for spacetime to be grounded in a non-spatiotemporal physics, then briefly consider the implications of causal spacetime emergence. Role-functionalist theories need a theory of realization, because they are theories intended to describe a domain of dependent properties such as mental or other special science properties. In traditional *flat* theories of realization, a functional property F is defined as the property of having some property that occupies role R (usually causal), and a property P realizes F by filling R. This theory presupposes that P and F are instantiated in the same object, and for that reason won't help us with the problem of empirical coherence. The properties whose realization we want to make sense of are spatiotemporal properties of local beables, but the fundamental objects in the putative realization base are, by hypothesis, not bearers of spatiotemporal properties. To account for the realization of spatiotemporal properties, we need a *dimensioned* theory of realization:

> A property-instance $F_\phi(x)$ is realized by properties and relations $P_1, \ldots, P_n$, $R_1, \ldots, R_m$ iff (i) x or its proper parts possess $P_1, \ldots, P_n$ and $R_1, \ldots, R_m$ in some combination; and (ii) x meets the specification ϕ definitive of F_ϕ in virtue of (i), but not *vice-versa*.[45]

[44] I don't claim here that role functionalism can handle the twin cases offered in Chalmers (2019). As a matter of fact I think it can, but I will defend that claim in future work.

[45] Gillett (2003). The 'not *vice-versa*' clause secures the asymmetry of realization. I have adapted Gillett's position so that F is no longer understood as causally individuated, and realization need not consist in bestowing powers. Realization of causally individuated properties by causal power bestowal is a special case of the definition above. For role-functional spatiotemporal properties as understood here, ϕ will be given by GR, and may well specify broadly causal roles, but I find it implausible that realizing GR spacetime could consist in causal power bestowal.

Now suppose local beables are composed of elements of the fundamental ontology—field quanta, modes of vibration of cosmic strings, or whatever they might turn out to be—that are not located in ordinary spacetime. The fundamental particulars don't have ordinary spatial or temporal properties and relations, but perhaps local beables can have their spatiotemporal properties in virtue of the *non*-spatiotemporal properties and relations of their proper parts. By way of illustration, consider again the Matrix. Suppose that what Matrix dwellers perceive as physical objects are composed of states of computer programs—the digital objects that combine to cause Matrix dwellers' experiences as of seeing an apple are parts of a *virtual* apple, whose fundamental nature is computational.[46] On a role-functionalist conception of spacetime, there's no obvious reason why Matrix virtual objects can't have computational properties and relations in virtue of which they satisfy the specifications that define spatiotemporal properties. If we can flesh out the details of the GR roles in a way that explains how they could be jointly occupied by computational properties of digital objects, then GR will be literally true in the Matrix, and not merely simulated, provided the roles in question are occupied there. There are of course many complications—not least the fact that the GR roles will include the way in which the mass-energy distribution curves spacetime, so realizing GR spacetime in a Matrix environment also requires that mass-energy is digitally realizable. Global role functionalism about fundamental physical properties, as we arguably find in a pure powers theory, would seem to have this consequence, but I cannot develop this proposal here for reasons of space.[47]

The position just sketched suggests that local beables need not instantiate the same locality properties as their fundamental proper parts, and that's just what is needed in order to make sense of the emergence of spacetime. Locality in GR spacetime, conceived as a role-functional property, is instantiated by non-fundamental local beables, with the fundamental proper parts of such things having different locality properties in the space that they inhabit.[48] Suppose global role functionalism about physical properties, and consider a Matrix world that is a qualitative physical duplicate of actuality. Within this Matrix's spacetime, the virtual Earth really is 149.6 million km from the virtual Sun, but they are composed of digital objects that are stored within the same supercomputer and (let's say) 1nm apart relative to our spacetime. If we take seriously the claim that the spatiotemporal properties of local beables are realized by the properties and

[46] Chalmers (2005); see Chalmers (2017) for more on the ontology of virtual reality.

[47] See Yates (2018).

[48] Le Bihan (2018) suggests that locality is relative to a structure, and distinguishes *fundamental* from *emergent* locality on that basis. The idea that a complex object could be located relative to GR spacetime, while its proper parts are not, is a consequence of allowing composition and realization to connect distinct locality structures.

relations of their fundamental proper parts, this kind of consequence should not be any more surprising than the idea that the mental is grounded in the non-mental.

The problem with this proposal is that classical mereology is, in Paul's terminology, *spatiotemporalist*:[49] we build complex local beables from more fundamental local beables, with the spatiotemporal properties of the whole dependent on the spatiotemporal properties and relations of the parts. In the Matrix example given above, it's hard to see how Matrix objects could be composed of computer program states unless those states themselves instantiate *spatiotemporal* properties and relations, in virtue of which they compose into complex objects. If we want to build things that are in spacetime from things that are not, we need an alternative mereology. Paul herself defends a non-spatiotemporalist version of mereological bundle theory, in which local beables are primitively restricted fusions of properties:

> Objects may have their locations in virtue of being fused with whatever location properties and relations there are that define the actual space of the world, and many objects will have a physical structure in virtue of having location properties and relations as parts of their fusions, or in virtue of being part of a larger fusion which has location properties and relations as parts. The character of the space might not be what we take the character of ordinary spacetime to be, but the structure of the space is generated by fusing qualitative properties with relevant properties and relations that define the space as determined by modern physics.[50]

Paul intends mereological bundle theory to be maximally flexible, so that objects, events, facts, states of affairs, etc. can be generated by bundling together fundamental qualitative and location properties, without the requirement that the location properties are spatiotemporal.[51] It's not entirely clear to me how to flesh out the details, but the general idea is that by abandoning spatiotemporalist mereology, we make room for an ontology in which fundamental things that are not spatiotemporally located can be proper parts of non-fundamental things that are. If we embrace a version of Paul's mereology with fundamental particulars as well as properties, we can maintain role functionalism about spatiotemporal properties without giving up on the dimensioned account of realization sketched above. The spatiotemporal properties of local beables will then be properties they have in virtue of the non-spatiotemporal properties and relations of their proper parts. A version of this claim may also be defensible on Paul's own view, although of course in that case the proper parts *just are* properties and relations.[52]

[49] Paul (2012). [50] Paul (2012, 242). [51] Paul (2012, 244–245).

[52] See Le Bihan (2018) for further discussion of the application of Paul's mereology to spacetime emergence.

What if GR spacetime is not grounded in a non-spatiotemporal fundamental structure at all, but emerges from such a structure *causally*? Suppose for the sake of argument that LQG is the correct approach to spacetime. If GR spacetime emerges causally from a spin foam, then there's no need to explain how the latter, complete with its disordered localities, could realize the former. Rather, the problem is to explain why the relevant non-localities disappear at low energies, and that's a problem for physicists, who need to supply a *causal* bridge principle in place of the metaphysical principle provided by realization.[53] As we saw in §6.2.2, if a LQG spin foam evolves into a low-energy state in which the disordered localities disappear, then its states may be suitable to occupy the roles of GR spatiotemporal properties. In this condition, I suggested, what we have is a spatiotemporal model of a theory, many of whose models will be non-spatiotemporal due to disordered locality. Role functionalism about spacetime is also consistent with this kind of possibility. If LQG states can occupy the GR roles, then they can realize a role-functionalist spacetime structure defined by those roles.

Now, if it turns out that the functional roles associated with *S*-concepts are indeed occupied by properties specifiable in the language of the fundamental theory, then we may well decide that a realizer functionalist approach is best. The issue there will depend, as it does (*mutatis mutandis*) in the philosophy of mind, on whether we think our concepts give us at least some epistemic access to a higher-order, multiply realizable spacetime structure, or simply enable us to refer to whatever occupies the relevant functional roles. However, if our goal is to develop a metaphysical framework that doesn't presuppose any particular theoretical developments in physics, then role functionalism is clearly preferable. It's not so clear, for example, that a string-theoretic target space of indeterminate geometry could evolve causally into GR spacetime, in the way we might imagine the non-localities of LQG disappearing at low energies. Role functionalism, if combined with suitable theories of realization and composition, leaves open that the fundamental ontology is synchronically non-spatiotemporal, and that gives us ample reason to work out the metaphysical details.

6.4 Conclusion

Working out how a non-spatiotemporal fundamental physical theory could entail propositions featuring spatiotemporal concepts is a problem for physicists, but metaphysicians can fruitfully think about the conditions under which this is possible. Antirealism, according to which spatiotemporal concepts pick out

[53] Wüthrich (2017); Lam and Wüthrich (2018).

elements of the fundamental ontology, is either untenable on conceptual grounds or unsuitable for addressing the problem at hand. The task then is to work out a version of realism—the theory that spatiotemporal concepts refer to derivative, non-fundamental properties and relations—that is both conceptually plausible and consistent with the broadest possible range of relations between the fundamental ontology and grounded spacetime. GR-based role spacetime functionalism allows spatiotemporal concepts to be scientific without denying their intuitive transparency, and allows GR spacetime to be synchronically grounded in a non-spatiotemporal fundamental structure. This is clearly not a fully worked-out theory, and many questions remain, for instance: (i) What exactly are the roles assigned to spatiotemporal properties in GR? (ii) Are those roles causal, nomic, structural...? (iii) What theory of realization is most appropriate for second-order properties defined by the roles in question? (iv) What metaphysics are required to allow for the composition of local beables by non-spatiotemporal proper parts? I leave these questions for future work.[54]

References

Block, N. (1978). Troubles with functionalism. In C. Savage (Ed.), *Perception and Cognition: Issues in the Foundations of Psychology*, pp. 261–325. Minneapolis, MN: University of Minnesota Press.

Chalmers, D. J. (2000). What is a neural correlate of consciousness? In T. Metzinger (Ed.), *Neural Correlates of Consciousness: Empirical and Conceptual Questions*, pp. 17–39. Cambridge, MA: MIT Press.

Chalmers, D. J. (2005). The matrix as metaphysics. In C. Grau (Ed.), *Philosophers Explore the Matrix*, pp. 132–176. New York: Oxford University Press.

Chalmers, D. J. (2012). *Constructing the World*. Oxford: Oxford University Press.

Chalmers, D. J. (2017). The virtual and the real. *Disputatio* 9(46), 309–352.

Chalmers, D. J. (2019). Three puzzles about spatial experience. In A. Pautz and D. Stoljar (Eds.), *Blockheads! Essays on Ned Block's Philosophy of Mind and Consciousness*, pp. 109–137. Cambridge, MA: MIT Press.

Churchland, P. M. (1981). Eliminative materialism and the propositional attitudes. *Journal of Philosophy* 78(2), 67–90.

[54] Versions of this chapter were presented at a seminar at the University of Geneva, at a symposium on spacetime emergence (organized by Baptiste Le Bihan) at ECAP9 in Munich, at a conference on metaphysics at the Federal University of Rio de Janeiro (organized by Guido Imaguire), and at a seminar at the University of Lisbon. Particular thanks to: Baptiste Le Bihan, David Chalmers, Karen Crowther, Natalia Deng, Nick Huggett, Bryan Pickel, Benjamin Schnieder, Chris Wüthrich, Elia Zardini, and two anonymous referees. Funded by the Fundação para a Ciência e a Tecnologia, award nos. IF/01736/2014 and PTDC/FER-HFC/30665/2017.

De Haro, S. (2019). The heuristic function of duality. *Synthese 196*(12), 5169–5203.

Gillett, C. (2003). The metaphysics of realization, multiple realizability, and the special sciences. *Journal of Philosophy 100*(11), 591–603.

Goff, P. (2017). *Consciousness and Fundamental Reality*. New York: Oxford University Press.

Healey, R. (2002). Can physics coherently deny the reality of time? *Royal Institute of Philosophy Supplements 50*, 293–316.

Huggett, N. (2017). Target space ≠ space. *Studies in History and Philosophy of Modern Physics 59*, 81–88.

Huggett, N. and C. Wüthrich (2013). Emergent spacetime and empirical (in)coherence. *Studies in History and Philosophy of Modern Physics 44*(3), 276–285.

Knox, E. (2019). Physical relativity from a functionalist perspective. *Studies in History and Philosophy of Modern Physics 67*, 118–124.

Kripke, S. A. (1972). Naming and necessity. In D. Davidson and G. Harman (Eds.), *Semantics of Natural Language*, pp. 253–355. Dordrecht: Reidel.

Lam, V. and C. Wüthrich (2018). Spacetime is as spacetime does. *Studies in History and Philosophy of Modern Physics 64*, 39–51.

Le Bihan, B. (2018). Priority monism beyond spacetime. *Metaphysica 19*(1), 95–111.

Le Bihan, B. and J. Read (2018). Duality and ontology. *Philosophy Compass 13*(12), e12555.

Lewis, D. (1972). Psychophysical and theoretical identifications. *Australasian Journal of Philosophy 50*(3), 249–258.

Lewis, D. K. (1994). Reduction of mind. In S. Guttenplan (Ed.), *A Companion to the Philosophy of Mind*, pp. 412–431. Oxford: Blackwell.

Lewis, D. K. (1995). Should a materialist believe in qualia? *Australasian Journal of Philosophy 73*(1), 140–144.

Lewis, D. K. (1999). *Papers in Metaphysics and Epistemology*. Cambridge: Cambridge University Press.

Markopoulou, F. and L. Smolin (2007). Disordered locality in loop quantum gravity states. *Classical and Quantum Gravity 24*(15), 3813–3814.

Matsubara, K. and L.-G. Johansson (2018). Spacetime in string theory: A conceptual clarification. *Journal for General Philosophy of Science 49*(3), 333–353.

Maudlin, T. W. (2007). Completeness, supervenience and ontology. *Journal of Physics A: Mathematical and Theoretical 40*(12), 3151–3171.

Ney, A. (2012). The status of our ordinary three dimensions in a quantum universe. *Noûs 46*(3), 525–560.

Ney, A. (2015). Fundamental physical ontologies and the constraint of empirical coherence: A defense of wave function realism. *Synthese 192*(10), 3105–3124.

Ney, A. (2017). Finding the world in the wave function: Some strategies for solving the macro-object problem. *Synthese*, 1–23. https://doi.org/10.1007/s11229-017-1349-4.

Paul, L. A. (2012). Building the world from its fundamental constituents. *Philosophical Studies 158*(2), 221–256.

Rovelli, C. and F. Vidotto (2014). *Covariant Loop Quantum Gravity: An Elementary Introduction to Quantum Gravity and Spinfoam Theory*. Cambridge: Cambridge University Press.

Schwitzgebel, E. (2019). Kant meets cyberpunk. *Disputatio*. https://doi.org/10.2478/disp-2019-0006.

Sklar, L. (1976). *Space, Time, and Spacetime*. Berkeley, CA: University of California Press.

Whittle, A. (2008). A functionalist theory of properties. *Philosophy and Phenomenological Research 77*(1), 59–82.

Wüthrich, C. (2017). Raiders of the lost spacetime. In D. Lehmkuhl, G. Schiemann, and E. Scholz (Eds.), *Towards a Theory of Spacetime Theories*, pp. 297–335. Basel: Birkhäuser.

Yates, D. (2012). Functionalism and the metaphysics of causal exclusion. *Philosopher's Imprint 12*(13), 809–841.

Yates, D. (2016). Demystifying emergence. *Ergo: An Open Access Journal of Philosophy 3*(31), 809–841.

Yates, D. (2018). Inverse functionalism and the individuation of powers. *Synthese 195*(10), 4525–4550.

7
Finding Space in a Non-Spatial World

David J. Chalmers

Abstract

What is the relation between space in the manifest image of perceptual experience and in the scientific image of physics? I will argue that we have moved from spatial primitivism (on which space is understood as a primitive conception that we are acquainted with) to spatial functionalism (on which space is picked out by its functional role). I investigate different forms of spatial functionalism on which the relevant roles are experiential (involving effects on our experience) and non-experiential (involving patterns of causal interactions). I draw connections to functionalism in the philosophy of mind, to Cartesian skepticism, and to recent literature on spacetime functionalism and emergent spacetime.

7.1 The Two Images

A familiar metaphor from Wilfrid Sellars (1962) says that there are two images that we confront and that we have to reconcile when doing philosophy. The manifest image is how things appear to be in everyday perception and thought. The scientific image is how things are according to science.

Numerous debates in philosophy try to reconcile these images. How do you reconcile the manifest picture of consciousness as it presents itself with what science is telling us about the brain and the mind? How do you reconcile the manifest image and the scientific image for free will, for meaning, for value, for color, or for time? Here I will confront that question for space.

How does space appear in the manifest image—the image presented to us relatively pre-theoretically in ordinary perception and thought? It appears that space is roughly Euclidean. Perception presents an absolute three-dimensional space evolving in time. Space arguably seems to be fundamental, a kind of fundamental container of all the things presented to us in the concrete world.

David J. Chalmers, *Finding Space in a Non-Spatial World* In: *Philosophy Beyond Spacetime: Implications from Quantum Gravity*. Edited by: Christian Wüthrich, Baptiste Le Bihan, and Nick Huggett, Oxford University Press. © David J. Chalmers 2021. DOI: 10.1093/oso/9780198844143.003.0007

Space as revealed in the scientific image, on the other hand, is non-Euclidean. There is a four-dimensional spacetime without absolute dimensions of space or time. On some current theories, spacetime may be non-fundamental.

What is the relationship between space in the manifest image and space in the scientific image? The relevant science here is physics. The idea I'll focus on is that in moving from the manifest image of space to the scientific image, we need to move from spatial primitivism, where we have a primitive conception of space, to spatial functionalism, where space is picked out as whatever plays the space role.

I argued for a functionalist view of space and time in my 2012 book *Constructing the World*. Here I'll motivate and develop spatial functionalism in more depth, using considerations from physics and also analogies with the metaphysics of color and with functionalism in the philosophy of mind. In the final two sections I'll connect these ideas to recent work in the philosophy of physics on spacetime functionalism and on emergent spacetime.

7.2 Primitivism and Functionalism about Color

I'll start with an analogy with the case of color. In the case of color, vision appears to present a world of colored objects. Out there in the environment now there are these deeply colored red chairs and there's a blue one and there's a white screen and so on. On the face of things, colors seem to be relatively primitive qualities of objects. The chair's surface is somehow imbued with redness. This chair's surface is somehow imbued with blueness. We seem to have some kind of fairly direct acquaintance with colors on the surface of these objects.

A metaphor that I like in making sense of the manifest image is to suppose that in the Garden of Eden everything was exactly as presented in the manifest image. Then Eden is a label for a world where everything is exactly as it seems to be.

How were colors in the Garden of Eden? It's arguable that the phenomenology of color experience presents us with a world of primitive color qualities on the surface of objects. In Eden, colors are just like this. Objects were imbued with primitive colors on their surfaces. We can call these colors Edenic redness, Edenic greenness, and so on. We stood in a direct acquaintance relationship to Edenic colors on the surface of objects.

Of course, when there's Eden there's a fall. Scientists ate from the tree of knowledge and discovered that we don't seem to live in Eden. People examined objects out there in the external world and they don't seem to have these primitive color qualities. All they have is some really complicated physical properties that reflect light in certain ways. Rather than having direct acquaintance with those things, there is a long and complex chain leading from these physical properties to the brain, involving light transmitted to the retina and electrical signals transmitted up the optic nerve.

Many people reacted to this by saying that colors don't really exist. The fact that people have that reaction is some evidence for the role of this primitivist conception of color in our ordinary conception of the world. Galileo, for example, reacted by saying objects in the world aren't really colored at all. Colors exist only in the mind and not out there in reality.

If you take this line, you have to say that apples aren't really red, and grass isn't really green. This is a kind of eliminativism about colors. In effect, colors are eliminated from the scientific image. A number of people have reacted this way, but over time it's been the less popular reaction. There's a more common reaction, which we can call compatibilism. Compatibilism says that even in light of the facts of color science and this fall from Eden, apples are still red, grass is still green. Even though we don't live in Eden, objects still have colors.

This is done in a variety of ways. Sometimes people identify colors with dispositions. Perhaps most common is to identify colors with certain complex physical properties such as surface reflectances and other physical properties such as radiation properties that affect light in appropriate ways. In effect, what's happening is that redness is now identified as that physical quality that plays a certain causal role. In particular, it is that physical quality that is causally responsible for our experiences of redness.

One can think of this metaphysical picture as a sort of imperfect realism. Imperfect realism is a picture that is very common in philosophy. It roughly says there are Xs, they aren't Xs exactly as they seem to be pre-theoretically, but they are still something that deserves well enough to be called Xs. In the case of colors, there are no perfect colors. There are no colors exactly as presented in experience, primitive Edenic qualities on the surface of objects. But there are still imperfect colors, properties that play the color role, for example in bringing about color experience.

What's going on here is that there has been a move from color primitivism, on which colors are regarded as primitive qualities, to color functionalism, on which colors are picked out as what play a certain causal role. In particular, colors are whatever is normally causally responsible for our color experience. As with other familiar varieties of functionalism about which we'll talk more soon, this view doesn't actually require that one be functionalist about the metaphysics of colors. It doesn't require that colors be functional properties. It's consistent with that view, but it's also consistent with the view that colors are physical properties that play that role.

This view tends to suggest functionalism about color concepts. Roughly, we're conceiving of colors as whatever causes color experience. In principle, one could run this without functionalism about color concepts by just appealing to a metasemantic causal theory of reference where we pick out whatever plays this role without our conceiving of colors that way. But I think reflection on how we go about picking out the colors in virtue of their serving as the basis of certain color appearances suggests that the strongest story involves functionalism about the relevant concepts.

When you have these Edenic colors and non-Edenic colors around, it's arguable there are also two kinds of concepts in the vicinity. There are at least two kinds of concept that somebody could have and arguably two kinds of concepts that we possess and that regulate our thought and talk about color. There's an Edenic concept, which we can call capital-R Red, for which primitivism and eliminativism are correct. We've got a concept of Edenic redness. It's a primitive concept, but nothing in the world turns out to satisfy it. We also have a non-Edenic concept, small-r red, for which functionalism and realism are correct. For small-r red, we just pick out colors as whatever causes the relevant experiences. It's a functional, non-primitive conception, but this is one for which realism is correct, because there are things that really play that role. Partly because of this, the functionalist concept is most useful in making sense of the world.

7.3 Primitivism and Functionalism about Space

The move from primitivism to structuralism or functionalism is a common picture throughout various areas of science when we try to reconcile a categorical manifest image with a more structural scientific image.

In my view, the same story applies to space. How was space in Eden? How is it presented to us pre-theoretically? The kind of space that we have an intuitive grasp on in our experience seems to have a certain categorical qualitative nature that serves as the container of everything we're presented with. In Eden, there were these perfect spatial properties—Euclidean distances, squares, and so on—all embedded in a primitive space and time. Then we ate from the tree of knowledge again. This time the tree of knowledge came in a number of stages.

The first really important step in the tree of knowledge for space was taken by Newton. Newtonian physics did much less to undermine the Edenic picture of space than corresponding steps around the same time did in the case of color. Newton suggests a fundamental Euclidean space, which at least seems consistent with the Edenic model of space.

More radical developments came later. Around the start of the last century, Einstein suggested a non-Euclidean space in which spatial properties are relative. Nothing is absolutely square. Things are just square relative to a reference frame. That's already a spatial fall from Eden. Using the intuitive concept of Edenic space that one gets directly from one's experience, I think it's arguable you can't really make sense of Edenic space being something that holds relatively rather than absolutely. In this move to relativize space, there's a fall from Eden going on.

It's arguable that the fall gets worse when it comes to quantum mechanics. The interpretation of quantum mechanics is controversial and contested, but one very common view is that what's fundamental in quantum mechanics are not things like the positions of particles, but wave functions. Wave functions are generally held to inhabit a high-dimensional configuration space. This is not a three- or four-dimensional space but roughly a 3n-dimensional space where n is the number of particles. That space corresponds to the space of possible configurations of those particles, with every configuration getting an amplitude. Instead of having fundamental 3-space, you have a fundamental 3n-space or infinite-dimensional space if the world is big enough. It starts to look like three-dimensional space isn't fundamental but arises derivatively from configuration space. At least, so it goes on many popular interpretations of quantum mechanics.

Things get even worse in string theory and other theories of quantum gravity, such as loop quantum gravity and causal set theory. These theories try to integrate quantum mechanics and relativity. Of course, that is much more speculative physics now. These theories all postulate various models on which space isn't fundamental at all. It's now common to suggest that space and possibly time are emergent in that they pop out at a derivative level of the theory. Nathan Seiberg (2006) talks about spacetime in string theory this way:

> [S]pacetime is likely to be an emergent concept. The fundamental formulation of the theory will not have spacetime and it will emerge as an approximate, classical concept which is valid only macroscopically. (Seiberg 2006, 20)

Getting even more speculative, there's digital physics, involving what people sometimes call the It from Bit hypothesis. That slogan is sometimes used for various different views. I'm concerned with views on which it turns out that underlying the apparent structure of physics including spatial structure is some computational level, perhaps something like a cellular automaton governed by an algorithm in a more fundamental space, which isn't our familiar 3-space—that the spatial structure of physics emerges from that. Stephen Wolfram (2003), Edward Fredkin (2000), and others have famously speculated along these lines.

Of course, quantum gravity theories and especially digital physics are admittedly speculative physics and the interpretation of quantum mechanics is controversial. But the kinds of interpretations of quantum mechanics that go in this direction are at least reasonably popular, and of course relativity is solid.

One reaction to all this is to say that space is an illusion if things are as presented that way in physics. We have the experience that things are located in 3-space when in fact they aren't. That's spatial eliminativism, analogous to color eliminativism.

It seems to require that all spatial experience is illusory and that all ordinary spatial claims are false. One can find a version of this claim in David Albert's 1996 paper on elementary quantum mechanics, though these days he accepts something closer to compatibilism (see Albert 1996, 2015). I take it that is a much more popular reaction in this case, as it is in the case of color. The compatibilist says that even if physics is this way, there is still space. It's just not exactly as we took it to be. It's going to be relative rather than absolute, and it's going to be a derivative rather than fundamental property. As Seiberg puts it, spacetime is an emergent concept. This is basically capturing the idea that space is non-fundamental and derivative.

The question that arises, if space is not fundamental, is how do we pick out the complex properties in this world that are spatial properties? I think the answer is very much analogous to the answer in the case of color. We pick out spatial properties as those properties that play the right role in the structure of the physical dynamics and in bringing about spatial experience. This is going to be a kind of imperfect realism. There are no perfect spatial properties, or primitive space exactly as presented in experience. But there are still imperfect spatial properties: complex properties that play the space role.

As in the case of color, I think we've moved from primitivism to functionalism. We started with a kind of intuitive spatial primitivism where space involves these primitive qualities that we're acquainted with, and everything is spread out in that primitive space. We've ended with a spatial functionalism. To use a familiar functionalist slogan: space is as space does. Or better: space is whatever plays the space role. As with color, spatial properties need not themselves be functional properties (as the first slogan suggests), but they are picked out as what plays the functional role. This is analogous to various other familiar sorts of functionalism where we pick out Xs as whatever play the X role. Colors are whatever plays the color role. It's very common to say for many theoretical terms in physics like mass: take mass to be whatever plays the mass role in the theory. For example, whatever properly resists acceleration and yields gravitational attraction in a certain way counts as mass.

On spatial functionalism, space is whatever plays the space role with respect to physical laws and/or experience. These are two distinct aspects to the role which I'll return to later on. Some varieties of spatial functionalism focus on the role in physical laws, and some focus on the connection to experience.

As in the case of color, I think there are at least two concepts of space. There's capital-S Space, Edenic space, for which primitivism and eliminativism are correct. We have this primitive concept of Edenic space, space as it was in the Garden of Eden. Relativity, quantum mechanics, and so on at least tend to strongly suggest that we don't live in that world. There is no absolute Edenic space. We've also got a non-Edenic concept, small-s space, for which functionalism and realism are correct. This concept picks out space as whatever plays the relevant role. Even in our post-fall world, something plays this role. As in the case of color, I think both concepts are in some sense perfectly coherent, but the functionalist concept is the most useful in making sense of the world we are confronted with. Again, this involves the move from intuitively having a categorical grip on the nature of space and spatiality to merely having a structural grasp of imperfect space, grasping it in virtue of the role it plays.

7.4 Spatial Functionalism and Explanatory Gaps

To some extent, spatial functionalism can be motivated even in physical frameworks where space is fundamental. Some bits of it can be motivated even in Newtonian and relativistic physics. One motivation comes from spatial Twin Earth cases. I discuss a number of these in 'Three Puzzles about Spatial Experience' (2019). There's Brad Thompson's (2010) case of Doubled Earth, suggesting that we don't have a categorical grasp of absolute size properties. Even in Newtonian physics, there's no absolute grasp of a length of one meter. That length is roughly picked out as whatever brings about certain kinds of experience. There's also the relativistic case of Lorentz Earth, suggesting that we don't have a categorical grasp of ordinary shape properties. Shape and relative size properties are also picked out in terms of the role they play.

Spatial functionalism becomes especially relevant and useful for making sense of frameworks where space is not fundamental, including some interpretations of quantum mechanics, quantum gravity, computational physics, and so on. In quantum mechanics, wave-function fundamentalism (e.g. Albert 2015 and Ney 2015) says all that's fundamental in quantum mechanics is the wave function defined over a configuration space evolving according to the Schrödinger equation. A number of popular interpretations of quantum mechanics have this form. Everett-style interpretations typically go this way. Collapse interpretations introduce extra dynamics with wave-function collapse, but many of them still say the wave function is all that's fundamental. The wave function is standardly taken to inhabit this high-dimensional configuration space. The question then arises: how do we recover familiar 3- or 4-space from high-dimensional configuration space?

The worry involves a kind of explanatory gap. How do you get from this massively high-dimensional wave function to the familiar three-dimensional space or four-dimensional spacetime that we're confronted with? It's not easy to find space of that kind in the wave function. On closer examination, I think spatial primitivism leaves an explanatory gap but spatial functionalism does not. This is an instance of a more general phenomenon involving functionalism and explanation.

In general, functionalism about a domain helps us to close explanatory gaps involving that domain. Familiarly, in the case of the mind–body problem, if you're a functionalist about the mental, you think that what it is to be in pain, for example, is to be in a state that plays the relevant causal role and that's what it means to be in pain. That helps you a lot in closing the gap from physical to mental. You can see roughly how in a physical world something would play that role. So if this analysis were true, you could see how a physical system could be in pain.

I don't endorse functionalism about pain myself, but I like the general structure. This is a version of the Canberra Plan that is familiar from the work of people like Frank Jackson (1998) and David Lewis (1994). To close an explanatory gap, say from the physical to the mental, or from the fundamental level to something non-fundamental, what you need is something like an a priori entailment. That way you can see why given that things were like this at the bottom level, they just have to be like this at the higher level. Somehow the relevant non-fundamental facts can be inferred from the fundamental facts. I don't want to put too much stress on the role of the a priori here but the rough idea is that we want some fairly transparent epistemic connection from the fundamental facts to the non-fundamental facts in order to close the explanatory gap.

Functionalism about the mental helps us see how that could be true. The picture from Lewis, Jackson, and others goes something like this.

1. A priori (pain is what plays the pain role).
2. A priori (if P, then C-fiber-firing plays the pain role).
3. A priori (if P, then x's C-fibers are firing).

————————————

4. A priori (if P, then x is in pain).

First, you hold it's a priori that pain is what plays the pain role. That's a functionalist analysis of pain. Then, let's say P represents relevant physical facts about the physical configuration of our world. The thought is that given these physical facts, one can deduce functional properties of the brain, and in particular deduce C-fiber firing is playing that role. It will also be deducible that in a certain individual, their C-fibers are firing. Putting all those together, from P it will be deducible that this individual is in pain.

This potentially closes the explanatory gap from physics to pain. It's controversial and I would reject the first step, but it's a nice general explanatory structure for locating pain in a physical world.

Likewise, functionalism about the spatial, if true, helps to locate space in a world where it is not fundamental. It closes the gap from the non-spatial to the spatial; for example, from a world that's fundamentally characterized at the wave-function level independent of space. If we have a concept of space for which it's a priori that space is what plays the space role, then I think you can plausibly make the case for a similar structure.

1. A priori (space is what plays the space role).
2. A priori (if P, then Q plays the space role).
3. A priori (if P, then x has Q).

———————————

4. A priori (if P, then x is in space).

Here P is the relevant fundamental physical configuration characterized independently of space, perhaps in quantum mechanical configuration space. Then if P is the case, one can deduce that some complex quantum mechanical state Q plays the space role. One can also deduce that some particular object x is in this state Q. Putting all those together, it's a priori that if P is the case, then I'm located in space, thereby closing the explanatory gap from the non-spatial to the spatial.

An austere version of this spatial functionalism project includes just fundamental physics (P) in the base. A less austere version allows the base to also include facts about experience. This can make it easier to locate space. It allows us to characterize the space role as not just a physical role but also a role in causing experiences. On this way of doing things, instead of closing the gap between physics and space we close the gap between physics plus experience and space. That will of course leave open the residual mind–body question of reducing experience, but it's fairly common to kick that problem down the road. I will look at both versions in what follows.

In the mind–body case all this gives rise to a complex structure of philosophical positions. We can put the issue by saying that these five views are inconsistent, so one of them has to go. Here phenomenal properties are properties characterizing what it is like to be conscious beings, and phenomenal concepts are our introspective concepts of those properties.

1. Physicalism (only physics is fundamental).
2. Phenomenal realism (things have phenomenal properties).
3. Phenomenal primitivism (phenomenal concepts are primitive).
4. Scrutability (no epistemic gap between fundamental and non-fundamental).
5. Epistemic gap (epistemic gap between physics and primitive phenomenal properties).

The first item in the quintet is physicalism: only physics is fundamental. The second is phenomenal realism: some things have phenomenal properties. That is, there are some beings (like us) for whom there is something it is like to be them that is the qualitative properties of experience. Phenomenal primitivism says that phenomenal concepts are primitive, meaning that they are not analyzable in non-phenomenal terms—and certainly that they are not functional concepts. Scrutability is roughly the claim that you can't have these deep epistemic gaps between the fundamental and the non-fundamental domain. The version of it I like is that the non-fundamental facts should be a priori entailed by the fundamental facts. Finally, there's the claim that there's an epistemic gap between physics and primitive phenomenal properties. If phenomenal properties can't be analyzed in non-phenomenal terms, there's an epistemic gap (and no a priori entailment) between the physical and the phenomenal.

Different people reject different claims here. Analytic functionalists reject phenomenal primitivism by saying that phenomenal concepts are functional concepts. Eliminativists reject phenomenal realism, saying that no one has phenomenal properties. So-called Type-B materialists reject scrutability, saying that the phenomenal can be ontologically grounded in the physical even if there is an epistemic gap. Some panpsychist physicalists reject the epistemic gap thesis by saying that physics builds in consciousness. I'm inclined to reject physicalism.

In the spatial case, we can lay out an analogous inconsistent quintet.

1. Wave-function fundamentalism (only the wave function is fundamental).
2. Spatial realism (things have spatial properties).
3. Spatial primitivism (spatial concepts are primitive).
4. Scrutability (no epistemic gap between fundamental and non-fundamental).
5. Epistemic gap (epistemic gap between wave function and primitive spatial properties).

The analog of physicalism here is wave-function fundamentalism. Then there's spatial realism, which says that things have spatial properties. Spatial primitivism says that spatial concepts are primitive—that is, that they are not analyzable in non-spatial terms, and certainly not in functional terms. Scrutability is the same principle as before, saying there is no epistemic gap between the fundamental and the non-fundamental. Finally, there is an epistemic gap between the wave function and spatial properties, at least as construed by the spatial primitivist.

Again, something has to go. The analog of my view in the case of the mind–body case would be to give up on wave-function fundamentalism and say that we need something else besides the wave function. Maybe we need some primitive particles in 3-space, maybe we need primitive space some other way, or something else.

In the mind–body case, my view is that phenomenal realism is more or less non-negotiable. When I say I am phenomenally conscious, in the ordinary sense of phenomenal consciousness, this is true. I also think it's extremely plausible that the ordinary phenomenal concepts used to express this claim are primitive concepts.

In the case of space, I don't think these things are non-negotiable. It's not a non-negotiable fact that things are in space in a primitive sense. We do have a primitive concept of Edenic space, analogous to our concept of phenomenal consciousness, but I think it's more palatable to deny that Edenic space exists than to deny that phenomenal consciousness exists. So I think the combination of spatial realism and spatial primitivism is something one can put much more pressure on than in the mind–body case. A spatial primitivist who takes our concept of space to be a concept of Edenic space should deny spatial realism. Likewise, a spatial functionalism can deny spatial primitivism and accept spatial realism. That's the combination I prefer.

An epistemological motivation here is also worth mentioning. Spatial functionalism has the potential to save the appearances in a certain sense when it comes to space. One sense of saving the appearances is saving exactly how things appear to be. In that case, to save the color appearances, what we really have to do would be to vindicate the Edenic picture of colors in the world. That we don't get to do, given color science. But we do vindicate the picture whereby we have these color experiences. Arguably, our having these experiences is the most fundamental data. If you regard those as the appearances, we can then construe colors as the things that bring about these experiences. Arguably, this saves the most fundamental data which are the data of our experience. Likewise, for spatial functionalism, as long as we're in a world where something is playing the right space role, it's arguably going to be the case that that will be a world that supports the same kind of spatial experiences.

Even though the external world may not be the Edenic world of our original conception, as long as things there function in the right way and bring about our experiences in the right way, then it will save the appearances. One thought here is that even if there's no Edenic space out there in the world, there'll be something out there which is playing at least a very broadly functionally isomorphic role in a functionally isomorphic world. A functionally isomorphic world, even if it's not Edenic, will at least have functional isomorphs of ordinary perceivers, like you and me as we standardly take ourselves to be. If you can motivate the case that in a world with such functional isomorphs of ordinary perceivers there will also be experiential isomorphs, people having corresponding experiences, then you've saved the experiences.

Now, admittedly, that last step is a strong step and interacts with controversial issues about the mind–body problem. But at least if you're inclined to think that what really plays a central role as a physical basis of experience is something about our functional dynamics, rather than, say, facts about our absolute spatial layout, then you'll be inclined to go this way. An opponent might try to make the case that you need Edenic space to have the kind of experiences that we have—for example, how could we have experiences as of Edenic space in a world without Edenic space?

That would be an interesting place to put some pressure. Still, I think this is a good prima facie case for some epistemological virtues of the view.

An even more austere picture is suggested by developing David Lewis's austere picture of Humean supervenience. On Lewis's picture, fundamentally there are just primitively distinct natural properties throughout spacetime. Lewis builds up the rest of the world from there by Ramsifying properties like mass and charge as what plays the mass role and what plays the charge role. He builds in space and time as fundamental, though. He never takes the further step of Ramsifying space and time via spatiotemporal functionalism, but it's a natural thing to do. That leads to an even more austere Humean supervenience that doesn't mention spacetime at the fundamental level. On this picture, a fundamental description just has logic, mathematics, and naturalness as its basic ideology. If one likes austere ontologies, I think this is an attractive picture of the world.

7.5 What Is the Space Role?

Now the question arises, what kind of functional analysis are we going to give of space? That is to say: when we analyze space as whatever plays the space role, what counts as the space role?

There are two basic options for functionally analyzing space (with many variants of each and also approaches that combine both). The first is a phenomenal analysis, where we pick out space in virtue of its role in causing relevant spatial experiences, very much analogous to the case of color. I've discussed that sort of analysis in 'Three Puzzles about Spatial Experience' (2019) and other work where it is motivated by the analogy with color and Twin Earth cases and the like. Here I want to focus on another option.

The second option is a non-phenomenal analysis where space is picked out as whatever plays the physical space role in the dynamics of the physical world. That line that doesn't have so much of a direct analog in the case of color. It looks like colors don't play such an interesting dynamical role independently of us in the physical world. Their role is mostly interesting only relative to brains organized in a certain way. But space is interesting and plays an important role independently of us, it seems. So it's possible to spell out key roles of space quite independently of conscious experience, in terms of what we might think of as the physical space role.

Why should we be concerned with that analysis? One reason is that this analysis might be more useful for materialists about the mind who want to reduce space to something more fundamental without presupposing any phenomenal or experiential notions. If we presuppose those notions, then that will leave your theory of space hostage to your theory of the mind, and at some point you may need to worry about the mind–body problem. More generally, it may be theoretically cleaner to make sense of the emergence of space in terms of physics alone. Many physicists and philosophers are naturally uncomfortable with giving minds a central role in making sense of space. Of course, space itself need not be mental on the phenomenal version of spatial functionalism. It's just that minds play a role in fixing reference to what counts as spatial, the same way that water appearances help fix reference to water. But even that may be more of a role than some people want to give.

What are the options for non-phenomenal functional analysis of space? A natural approach is to use the Ramsey-Carnap-Lewis method for analyzing theoretical terms. We start with a theory of space, including our relatively pre-theoretical folk theory and/or our relatively sophisticated scientific theories. We list the central principles concerning space in the theory, or the platitudes, as they're sometimes called. We regiment these principles concerning space into a space role. Then we identify space with whatever fills the role. For mass, the mass role might be to resist acceleration and undergo gravitational attraction in certain ways. Then mass is whatever plays that role. That is, it is whatever resists acceleration and is subject to gravitation and so on.

What are the relevant roles in the case of space? I think there are a few roles. First, there are what we might call structural roles characterizing the broad abstract structure of space. Pre-theoretically at least, space seems to be three-dimensional with one corresponding dimension of time. There seem to be locations in space with distances between them. It seems that objects take up space, they're located at various locations in space at least at times, and they thereby have properties like shapes, sizes, and relative distances. One can go on to spell out broad structural characterizations of space in these terms.

Second, there are phenomenal roles, which I've talked about already. There are various things we might think of as the phenomenal platitudes: square things normally look square, small things normally look small, distant things normally look distant, and so on. There's some room for illusions where we get things wrong but the thought is, by and large, things are roughly the way they seem to be. This is basically a role connecting spatial properties of objects with certain kinds of spatial experiences which we can then perhaps use to characterize objects as what play that role with respect to experience.

Third, there are what we might call folk physics roles. There's a sort of pre-theoretical physics that's familiar from being an agent in the world before you do a lot of science. Psychologists have studied folk physics quite a lot. According to folk physics, our bodies are located in space, and there are many familiar objects like bodies, rocks, tables, trees with familiar shapes, sizes, relative locations. Their spatial properties evolve according to certain familiar principles. You throw them up in the air and they come down. If something is moving in a certain direction, it keeps moving in that direction, although maybe it gradually slows down according to folk physics.

Fourth, there are scientific roles. Science tells us that space plays such and such roles. It roughly obeys principes of Newtonian mechanics and Newtonian gravitation in ordinary contexts. More precisely, it obeys principles of relativity theory, quantum mechanics, and so on. There's a question of how much of this science one wants to and one is entitled to bring in to one's analysis of space here. The basic worry is that the fact that space plays these roles is only discovered by science and so seems to be a posteriori. So if one is analyzing our ordinary pre-science concept of space, these roles don't seem to be part of the concept.

At the same time, we have post-science concepts of space, such as the concept of Newtonian space, for which the roles of Newtonian mechanics are absolutely central. It's arguably a priori of Newtonian space that it plays the Newtonian roles. In a similar way there's a relativistic concept of spacetime, and it's arguably a priori of relativistic spacetime that it plays the relativistic roles. For explaining why there is something approximately like Newtonian space in a relativistic context, you want to apply the Newtonian concept. You show that in relativity, something approximately plays the Newtonian space role under certain conditions. To explain why there is something like relativistic spacetime in string theory, you want to apply the relativistic concept. You show that in string theory there is something that at least approximately plays the relativistic spacetime role under certain conditions.

In general, to pick out space in some later theory (say, string theory), we will typically want at least a relatively a priori analysis that does not presuppose that theory. But it might presuppose some earlier theory such as relativity. You're not exactly vindicating the manifest image in terms of the scientific image, but you're vindicating at least a relatively manifest image (say, the image of Newtonian space) in light of a relatively scientific image (say, the image of relativistic spacetime).

So we have different concepts of space with different analyses. I am especially interested in our ordinary concepts of space, so I'm interested in relatively pre-theoretical roles. Now, you might say, all these functional analyses of space are bad analyses of the pre-theoretical concept, just as the functional analysis of pain is a bad analysis. Many people including me think the functional analysis of pain is bad because you can conceive of all those roles being played without pain, and maybe of pain without those roles being played. Likewise, you might say, 'I can conceive of a space which plays none of these roles and maybe I conceive of these roles being played without space'.

I'm sympathetic with this point to a degree. I think, yes, one can conceive of space that plays none of those roles. Perhaps there could be a really weird space that evolves by completely different dynamics that's not connected to experience in these ways and so on. In these intuitions I think I'm basically invoking the primitive or Edenic conception of space, which I think is not a functionally analyzable concept. Maybe you think you can conceive of a non-primitive space that's not the Edenic conception and still doesn't play these dynamic or folk physics or scientific roles. Arguing over this will involve some litigation of cases. My guess is that one will be invoking either the primitive or the phenomenal conception at that point.

Our pre-theoretical conception of space involves these primitive elements on my view, but it also involves functional elements. These might help to ground a functional concept of space that's both pre-theoretical and non-phenomenal. Certain dynamic roles from folk physics may be especially crucial, especially certain roles in motion and in interaction. A central principle of folk physics says that objects move continuously through space. Another says that objects interact with nearby objects in space. To a first approximation, there's no discontinuous motion and there's no action at a distance. These are commonly taken at least to be principles of folk physics. Maybe they are violated here and there in scientific physics. But even in scientific physics, action at a distance is heavily constrained and there are many forms of action and motion that are local. That is, there seems to be less action at a distance and less motion at a distance, so that distance still has some connection to the possibility of interaction and the character of motion.

These principles are a central part of the folk conception of space. If we apply the Ramsey method, what we have to do is turn the principles around and identify distance with what plays the distance role. Instead of saying there's no action at a distance, we turn this around and we say, 'Distance is what there's no action at'. This is a nice slogan that goes back at least to Brian Cantwell Smith in his 1996 book *On the Origin of Objects*; here he gives something like this functional analysis of distance in a very different context.

In a similar way, we can use the principles about continuous motion to say, 'Distance is what there's no motion at', or equivalently, that location is something that changes continuously. Or if you want to do this in light of the science of distance, we can say that distance is what there's less action at and less motion at.

There's a very nice statement of this kind of distance functionalism in Gregg Rosenberg's 2004 book *A Place for Consciousness*, which advocated panpsychism about the mind–body problem as well as a revisionary view of causation in the physical world. Along the way Rosenberg also gave what he called a kind of causal analysis of spacetime. He put this point wonderfully by saying, 'There's a causality condition on locality, not a locality condition on causality' (157). A locality condition on causality says, it turns out, that causation only works in a local way. A causality condition on locality says that what it is to be close is to have the possibility of certain sorts of causal interaction. These claims may have to be weakened a little to accommodate a certain amount of non-local causation, but this is consistent with an underlying conceptual connection between the ideas.

We can think of this as the interactive conception of space. It's perhaps just one strand in our ordinary conception of space, but it can play a key role in trying to make sense of one aspect of the manifest image of space. This is a really central aspect in light of the scientific image of space.

It's arguable that this interactive notion of space and its role in motion and interaction and so on corresponds to a more generic notion of space. We've got this notion of a space in a general sense, which transcends physics. People talk about social spaces and virtual spaces and administrative spaces and teaching spaces and cyberspace. There are spaces for everything these days. Something in this generic notion of space seems to be tied to thinking of space as a generic realm in which a sort of motion takes place and in which interaction takes place. There are locations in these spaces and locations are close in a space when things move and interact between those locations.

Perhaps at least when combined with some further constraints, this pre-theoretical notion of interactive space can at least help to ground a reduction of physical space. Say, for example, if in the quantum mechanical wave function one can find a three- or four-dimensional metric of relations so that motion is relatively continuous and interaction is relatively constrained by that particular metric, that could at least help to ground a reduction of 3-space or 4-space in the world of quantum mechanics.

One may well need further constraints. Just what constraints on a functional analysis of space does one need to ground a reduction in quantum mechanics and string theory and computational physics? The details I take to be very much an open question.

One route that may help is a phenomenal analysis of space. Here we can pick out spatial properties in a quantum-mechanical or string-theoretic world as what causes spatial experiences in that world. As long as the wave function or the string-theoretic structure does bring about our experiences, then there ought to be something systematic to say about what quantum-mechanical or string-theoretic properties cause these experiences. Of course, there's the worry about giving the mind a role in fixing reference to physics, but I think this happens a lot. One still needs to worry about how spatial experiences can arise in a fundamentally non-spatial world. That problem will become part of the mind–body problem, perhaps making the problem a little harder. For a non-reductionist about the phenomenal, the solution will come down to finding laws that connect certain non-spatial properties in brains and the like to spatial experiences. Anyway, there's at least a program for finding the causes of our spatial experiences in a fundamentally non-spatial world, and I think it's reasonable to be optimistic about finding things that can play that role.

If you want a completely non-phenomenal functional analysis of space, things are trickier. It's perhaps easier for theoretical concepts of space where theories give a fairly precise role. As before, one can try to find something that plays roughly the Newtonian role or the relativistic role for space in a quantum-mechanical world or a string-theoretic world without space at the fundamental level. This is highly nontrivial but can be pulled off if there's the right sort of fit between the theories. In delivering something akin to Newtonian space or relativistic spacetime, it may not quite deliver exactly the original pre-theoretical manifest image, but it still delivers some relatively pre-theoretical image.

The hardest case is grounding a reduction of space with a non-phenomenal analysis of the pre-theoretical concept. It may be that the roles in interaction and motion that I've been gesturing toward can be used to deliver something even closer to the manifest image of space, picking out underlying properties in a quantum-mechanical or string-theoretic world as the grounds of motion and interaction. Or perhaps one can bring in other roles of space from pre-theoretical folk physics and see what plays them in the relevant physical theories. I take that to be an open question which is going to involve a lot of detailed work in the philosophy of physics combined with philosophical and/or psychological analyses of our concepts.

7.6 Virtual Reality and Cartesian Skepticism

Another interesting case is finding space within virtual reality. In other work I've argued that spatial functionalism applies there as well.[1] In virtual reality systems, there is typically something playing the space role. So if we're inhabiting virtual reality, we can identify space with those digital properties that play the space role.

This has some bearing on Cartesian skepticism. A Cartesian skeptic typically suggests that for all we know, we're in something like a Matrix scenario where nothing that we seem to experience is real. But if spatial functionalism is right, then if we're in a Matrix scenario, space is quite real. At least, space is no more illusory than it is in quantum gravity. And once the space around us is real, it's a short step to the objects in it being real too.

On this way of looking at things, if we turn out to be living in the Matrix, that's just another stage in the fall from Eden analogous to the Galilean and Einsteinian falls from Eden. After Galileo, red was identified as a reflective property. After Einstein, squareness was identified as a relative property. After the Matrix, squareness was identified as a virtual property. We've re-made our conception of it. We don't need to say things aren't square. We just say that squareness is not exactly what we took it to be.

[1] See Chalmers (2017) for virtual reality and Chalmers (2018) for applications to skepticism.

There's still an intuition that the Matrix scenario is an error scenario, one where people get things wrong. I have the intuition. But I think this is best explained by our using Edenic concepts in our intuition. In a Matrix scenario, there is no Edenic color or Edenic space. The Edenic content of our experiences is incorrect, and our experiences are not perfectly veridical. But if this makes the Matrix a skeptical scenario, then the same goes for quantum mechanics, relativity, and so on. I've argued that this is too high a standard. By the lower standard where quantum mechanics comes out looking okay, so does the Matrix. If so, spatial functionalism has some nontrivial purchase in addressing this sort of external-world skepticism.

This is an instance of a general phenomenon: primitivism about Xs tends to open the door to skepticism about Xs, while functionalism about Xs makes skepticism about Xs harder to get off the ground. That applies to consciousness, to color, to space, and to the physical world in general.

7.7 Spatial Functionalism and Spacetime Functionalism

In the last few years, there has been a small explosion of work on 'spacetime functionalism' in the philosophy of physics. Eleanor Knox (2014, 2019) has put forward spacetime functionalism as an interpretation of spacetime in general relativity. Vincent Lam and Christian Wüthrich (2018) have applied spacetime functionalism to analyze the emergence of spacetime in theories of quantum gravity with a fundamental non-spatiotemporal base. David Albert (2015, 2019) has developed his earlier work that takes a functionalist approach to space in quantum mechanics. Marco Dees (2015) developed a 'causal theory of spacetime' on which all geometric properties are non-fundamental and grounded in causal facts at the fundamental level.[2]

How is spacetime functionalism related to the spatial functionalism that I discuss in this chapter?[3] Both are versions of spatiotemporal functionalism: the broad family of functionalist views of space, time, and spacetime. There are many different views within this class, however. In this penultimate section I

[2] Dees died tragically in a rock-climbing accident in July 2018. His ideas on the causal approach to spacetime remain unpublished, but his 2015 Rutgers Ph.D. dissertation with a chapter on the topic is available online.

[3] In my own work, I first discussed space and time in a broadly functionalist spirit in 'The Matrix as Metaphysics' (2005) and 'Perception and the Fall from Eden' (2006). I first talked of 'spatial functionalism' and 'spatiotemporal functionalism' in *Constructing the World* (2012). These ideas were developed in four Jean Nicod lectures in 2015, which roughly correspond to 'The Virtual and the Real' (2017), 'Three Puzzles about Spatial Experience' (2019), 'Structuralism as a Response to Skepticism' (2018), and this chapter. Of course, one can find functionalist ideas about space in a good deal of earlier work. In *Constructing the World* I argue that there are elements of spatial functionalism in Berkeley and Hume. I was also greatly influenced by the work of Gregg Rosenberg (2004) and Brad Thompson (2003, 2010).

will map out the relationship between my spatial functionalism (especially my version) and spacetime functionalism (especially Knox's version), considering other functionalist views along the way. In the final section I will connect all this to issues about the emergence of spacetime.

First, the similarities. Both spatial functionalism (as I present it) and spacetime functionalism (as Knox presents it) are theories of spatiotemporal *concepts*. I say that the concept of space is the concept of whatever plays the space role. Knox says the concept of spacetime is the concept of whatever plays the spacetime role. So both are forms of *realizer functionalism*: space or spacetime is identified with what plays (or realizes) the space role or the spacetime. This contrasts with *role functionalism* which identifies space or spacetime with the playing of the role. Correspondingly, these are not forms of metaphysical functionalism, on which spatial properties or spacetime properties are identified with functional properties (such as the property of playing a certain role). Rather, they are forms of conceptual functionalism, where the concepts of space and spacetime are analyzed as the concept of whatever plays the role, so that space and spacetime themselves can be metaphysically identified with what plays the role.

Second, the differences. The most obvious difference is that spatial functionalism is a theory of space while spacetime functionalism is a theory of spacetime. One might think this is just a matter of considering time as well as space, in which case there would be little substantial difference, especially as I discuss 'spatiotemporal functionalism' in *Constructing the World* and also advocate functionalism about time. However, there is a more significant underlying difference. Spacetime as understood here is an essentially theoretical concept, one that emerges especially from the general theory of relativity. This concept can be applied to other theories, including Newtonian theories and theories of quantum gravity, but the concept itself did not exist (at least in the relevant sense) prior to relativity. Space is not an essentially theoretical concept. It can be used theoretically, of course. But there is also an ordinary concept of space used by ordinary people before and after the advent of theories of space.

The biggest difference between spacetime functionalism and my spatial functionalism is that the former is primarily an account of the theoretical concept of spacetime, while the latter is primarily an account of the ordinary concept of space. To a first approximation, spacetime functionalism is concerned with the status of spacetime as understood by physicists. I am concerned with the status of space as understood by ordinary people—though perhaps in light of developments in physics. That is the point of starting with Sellars' two images. The project is to see how ordinary space can be located within our scientific conception of the world.

This difference mirrors a familiar distinction in the philosophy of mind. This distinction is sometimes cast as the distinction between *analytic functionalism*, which offers armchair conceptual analyses of mental concepts in functional terms, and *psychofunctionalism*, which offers scientific analysis of mental concepts in functional terms. I don't think this is the best way to put things, at least for our purposes. This model suggests that analytic functionalism and psychofunctionalism are rival theories of ordinary mental concepts (and perhaps of their referents, ordinary mental properties). What we are concerned with here are compatible theories of two different things: ordinary concepts and theoretical concepts (and perhaps of their referents, ordinary properties and theoretical properties).

The distinction between *commonsense functionalism* and *theoretical functionalism* is more apt for our purposes.[4] The former analyzes the ordinary commonsense term of belief, memory, or whatever, as used by ordinary people. The latter analyzes the theoretical term, as used by theorists. Both may well be forms of analytic functionalism, and they need not be in conflict with one another. One gives a conceptual analysis of an ordinary concept, the other gives a conceptual analysis of a theoretical concept. I think that much of what goes under the label psychofunctionalism in the philosophy of mind is best understood as theoretical functionalism.

Using this framework, spacetime functionalism is clearly a form of theoretical functionalism, as the concept of spacetime is a theoretical concept. Spatial functionalism comes in both commonsense and theoretical varieties, as there are both ordinary and theoretical concepts of space. I have touched on both of these in my work on spatial functionalism (for functionalism about the theoretical concept, see pp. 332 and 418–419 of *Constructing the World*, as well as section 7.5 above), but my primary focus has been on the ordinary concept and on commonsense functionalism.

As a result, it is no surprise that the analyses delivered by my spatial functionalism and spacetime functionalism look very different. Spacetime functionalism uses a spacetime role tied heavily to the role of space within theories such as general relativity. My spatial functionalism has focused on a space role tied more to folk physics, including the role of space in appearance, motion, and interaction.

This leads to one striking difference, already touched on in the last section: spatial functionalism often involves phenomenal roles (roles in causing spatial experience), whereas spacetime functionalism typically abjures those roles. I don't think this is an essential difference. As I discussed in the last section, there are versions of ordinary spatial functionalism that invoke only non-phenomenal roles such as roles in motion and interaction. In principle there are also versions of theoretical spacetime functionalism that invoke phenomenal roles, since a

[4] The labels 'commonsense' and 'theoretical' are not perfectly apt, since analyses of an ordinary concept can be theoretical (through a folk theory) and need not obviously be tied to commonsense. 'Quotidian' vs. 'scientific' functionalism might be better. But there are enough labels and distinctions floating around that I do not want to introduce more.

connection to observation can be seen as part of the relevant theory. Still, it is easy to see why phenomenal roles may be more central to the ordinary concept than to the theoretical concept.

Another difference between the projects is that the literature on spatial functionalism has focused mainly on specific spatial concepts such as distance, shape, and size. By contrast, as Baker (2019) and Yates (this volume) observe, the spacetime functionalism literature has focused largely on the overarching concept of *being a spacetime* and not on the specific concepts. Baker suggests that an analysis of the overarching concept might be leveraged into an analysis of specific concepts if we appeal to principles of general relativity whereby a spacetime structure fixes an underlying metric. However, these principles will then have to be built into the analysis of the specific concepts, and it is not clear that the resulting concepts will be applicable outside the context of general relativity. So at least more work needs to be done to analyze specific spatiotemporal concepts in the framework of spacetime functionalism.

What is the correct analysis of the overarching theoretical concept of spacetime? In work on spacetime functionalism, there has been some debate over just what counts as the spacetime role. Knox focuses on an inertial role where spacetime is whatever fixes a structure of local inertial frames. James Read and Tushar Menon (2019) focus on an operational role tied to the way spacetime affects rods and clocks. David Baker suggests a cluster concept with many different components of the role. The argument between them can resemble old-school conceptual analysis. For example, Baker and Read and Menon call Knox's analysis into question by offering counterexamples in the form of theories that seem to have a spacetime but that do not meet her analysis. It is hard to see what settles the question of which analysis is correct. Does this turn on our intuitions about which theories and scenarios are best described as involving 'spacetime', or on psychological and linguistic facts about use of the key terms by physicists? It is not obvious why we should care deeply about these facts except as a matter of the sociology of science. And in practice, even participants in these debates (including Knox) are happy to depart from these intuitions when it suits their purposes.

My own metaphilosophy here dictates *conceptual pluralism* (see Chalmers 2011). There are many interesting concepts in the vicinity of most important terms, and what matters most in choosing among them is not linguistic usage but explanatory roles. 'Space' and 'spacetime' are no exceptions here. In particular, there may be more than one theoretical concept of spacetime. The reason is that there are many different theoretical roles that spacetime plays, and for each role there is in principle a concept of spacetime as what plays that role. Some of these may come closer to actual usage of 'spacetime' than others, but this may just reflect contingent facts about usage. More important, perhaps, is that different concepts of space will be important for different purposes. For example, Read and Menon argue that Knox's inertial concept is more useful than Baker's cluster concept precisely because it 'has the capacity to be put to novel interpretative work' in classifying what counts as spacetime under various theories. On the other hand, the operational concept may more closely track the way that spacetime is measured. And a phenomenal concept of spacetime, which connects spacetime in the theory not just to rods and clocks but to conscious observation, may help bridge the gap between the scientific image and the manifest image.

Some degree of conceptual pluralism may also be true for the ordinary concept of space. One strand in it involves a primitive concept of Edenic space. Another strand involves the role of space in being about spatial experience. Still another involves the role of space in motion and interaction. I think all of these strands are present in the ordinary conception of space. In pre-scientific days, the primitivist concept may have been the most useful, but even then the functionalist strands were present. After developments in science, the functionalist concepts became more useful, because they manage to pick out something important in the external world. As always, all of the concepts are available, but some are more useful than others. It is the functional concepts that best serve to locate the key elements of our pre-theoretical conception of space (the manifest image) in light of our best scientific theories (the scientific image).

7.8 The Emergence of Spacetime

One of the central motivations for spatiotemporal functionalism is the one suggested by my title: to help locate space in a world in which space is not fundamental. In 'The Matrix as Metaphysics' (2005, section 3 and note 14), I used the possibility that space emerges from a deeper level as part of my argument for the broadly functionalist view of space articulated there. In recent years, this motivation has received increased attention. Discussion of 'emergent spacetime' has exploded, driven largely by theories of quantum gravity—including string theory and loop quantum gravity—in which spacetime may not appear on the fundamental level. The relevant notion of emergence is not the philosopher's metaphysically demanding notion of strong emergence but the scientist's less demanding notion of weak emergence (see Le Bihan 2018, Baron 2019, and Wilson, this volume, for discussion). The key thesis is that spacetime exists at a non-fundamental level and is grounded in a fundamental level which is non-spatiotemporal. A number of philosophers (including Lam and Wüthrich 2018 as well as Le Bihan and Baron) have discussed the idea that spacetime functionalism can play a role in this grounding. In this final section I will draw some connections between my approach and this recent literature, aiming to get clearer about the prospects for functionalist approaches in vindicating emergent spacetime.

The idea of emergent spacetime has sometimes been received with skepticism.[5] Underlying much of this skepticism is the issue I discussed in section 7.4—that there seems to be an explanatory gap between the non-spatiotemporal and the spatiotemporal. Whatever emerges from the non-spatiotemporal could not genuinely be *spacetime*. A number of theorists have used analogies with explanatory gaps in the philosophy of mind to flesh out this point.

Baptiste Le Bihan (2019) articulates the worry by suggesting that there is a 'hard problem of spacetime' that is not addressed by standard approaches. Perhaps these approaches can solve the 'easy problems' of recovering some of the mathematical structure of spacetime theories from a non-spatiotemporal basis. But they do not solve the 'hard problem' of explaining 'spacetime qualia'. By this, Le Bihan means certain special qualities of spatiality and temporality that we associate with spacetime. The word 'qualia' is perhaps misleading, as it strongly suggests mental qualities of conscious experience, and Le Bihan is clear that he does not mean this. So I will speak simply of *spacetime qualities* instead.

In my framework, the hard problem of spacetime is the task of explaining Edenic space and time: space and time as they appear to be in our experience. There is a problem of explaining how objects in the world could be laid out in a genuine space of the sort that there seems to be in our spatial experience of the world, with time genuinely passing as it seems to be in our temporal experience. All that can seem to require special qualities of Spatiality and Temporality. I would identify these qualities with Edenic spatiality and temporality.[6]

The hard problem of spacetime is analogous to what Alex Byrne (2006) has called 'the hard problem of color': explain how things in the world could have genuine colors like red or green, as they appear to be in our experience, in a world where colors are not fundamental. In my framework, Byrne's problem is the problem of explaining how physical objects could have Edenic colors: the primitive sensory qualities such as Redness and Greenness that color experience seems to present objects as having. There seems to be a gap between the non-color-involving properties of the world that physics gives us and genuine Color. Byrne thinks that the hard problem of color is at least as important as the hard problem of consciousness.

The most common reaction to the hard problem of color is to deny that objects have the special qualities such as Edenic redness and greenness that they seem to

[5] For articulations of this skepticism, see Maudlin (2007), Le Bihan (2018, 2019) and others.

[6] One difference is that Le Bihan does not question whether general relativity can accommodate his spacetime qualities (perhaps because spacetime is fundamental in general relativity), while I say that it cannot accommodate Edenic space and time (because space and time are not absolute in general relativity). On my picture, relativistic spacetime is fundamental but not Edenic. One could perhaps reconcile these pictures by holding that the relativistic fall from Eden is only partial and that there are some aspects of Edenic spacetime that are compatible with a relativistic world.

have. Those qualities do not exist in ordinary physical objects. At best they exist or are represented in the mind. At some level perception may present apples as having these special qualities, but to this extent perception involves an illusion. We solve the hard problem of color by removing the special qualities from the world and putting them into the mind, where they become an aspect of the hard problem of consciousness. In the external world, all we have are small-c colors, understood functionally as properties that play the color roles.

If my analysis of the explanatory gap for space (in section 7.4) is correct, we should respond to Le Bihan's hard problem of spacetime in the same way. That is, we should deny that the physical world has the special qualities of Edenic spatiality or temporality. At best these qualities exist or are represented in the mind. At some level perception may present the physical world as having these special qualities, but to this extent perception involves an illusion. We solve the hard problem of spacetime by removing the special qualities from the world and putting them into the mind, where they become an aspect of the hard problem of consciousness. In the physical world, all we have is small-s space and small-t time, understood functionally as properties that play the space role and the time role.

In principle, an analogous view is available for the hard problem of consciousness. We could deny that consciousness has the special qualities it seems to have, and say this is a sort of introspective illusion. This is the move made by illusionists about consciousness such as Daniel Dennett and Keith Frankish. Most people find this view much harder to believe than the corresponding view for color and space, in part because the existence of the relevant qualities of consciousness seems to be something of an introspective datum. This apparent datum is one of the things that makes the hard problem of consciousness hard. By contrast, the idea that physical objects have special qualities of color, space, and time does not seem to be a datum in the same way. As a result, the hard problems of color and of spacetime are easier to dismiss than the problem of consciousness.

If I am right, then as long as quantum gravity theories can plausibly say there is spacetime but not Spacetime, they can avoid the hard problem of spacetime. And as I argued in section 7.5, it is arguable that functionally construed spacetime is enough to save the appearances of our sensory experience, at least if we accept that functional isomorphs of ordinary perceivers will have the same sensory experiences. If so, then spatiotemporal functionalism holds out the promise of an empirically and philosophically adequate theory of spacetime.

In a similar vein, Lam and Wüthrich (2018) have raised the problem of 'spacetime zombies' for spacetime functionalism. It seems conceivable that something could play the spacetime roles perfectly, without there being genuine spacetime. If so, being spacetime is not just a matter of playing the spacetime role.

One can treat this suggestion as I suggested in section 7.5, saying that the zombie intuitions arise primarily for Edenic or primitive space. That is, Spacetime zombies may well be possible, but not spacetime zombies. If we can plausibly deny that our world has Spacetime, this is no problem. It is also worth noting that realizer functionalism can accommodate a certain sort of spacetime zombie even without invoking Edenic space. We might imagine a perfect functional simulation of our universe in a computer—but the simulation will not have the spatial properties of the universe (e.g. it may be much smaller!). A realizer functionalist will explain this by noting that spacetime is identified with the *actual* realizer of the spacetime role in our normal environment. The simulation has a different realizer, which is not spacetime. This sort of spacetime zombie is quite consistent with the functionalist approach.

Sam Baron (2019) and David Yates (this volume) present related dilemmas for the realizer functionalist approach to spacetime emergence. The first horn of the dilemma focuses on the possibility that the realizers are themselves fundamental properties in a quantum gravity theory. In this case, these spatiotemporal properties will themselves be fundamental properties and we will not have a genuine form of spacetime emergence. On the second horn, the realizers are non-fundamental properties. But then we need a substantial account of how the non-fundamental realizer properties are grounded in fundamental non-spatiotemporal properties. Spacetime functionalism tells us how the realizer properties ground spacetime, but it tells us nothing about how these properties are grounded themselves.

I favor the second horn. How might the realizer properties be grounded in the non-spatiotemporal? One strategy appeals to mereological composition: fundamental non-spatiotemporal entities stand in a part–whole relation to spatiotemporal realizer properties. Baron argues that this cannot work, as there are conceptual connections between parthood and spatiotemporal location. A realizer functionalist might question this and appeal to a more general notion of composition that is not tied to location. Or perhaps better, they may appeal to non-mereological grounding relations. For example, fundamental causal powers in a non-spatiotemporal ontology may ground higher-level causal powers which may themselves serve as realizers for the functional roles associated with spacetime. Or high-level structural properties may be mathematically abstracted from low-level structural properties. The precise grounding of spacetime in the non-spatiotemporal remains an important open question, but there are many options available here.

This leaves open the question: what sort of functionalist analysis of space or spacetime is most promising for vindicating the emergence of spacetime in quantum gravity theories? This depends in large part on the details of the theories. The general shape of the project is clear: focus on one or more roles played by spacetime in general relativity (say), and make the case that under a certain theory of quantum gravity, a certain non-fundamental structure S will play that role. If that role is our criterion for counting as spacetime, then S will count as spacetime by this criterion.

Ideally, under quantum gravity, there will be some non-fundamental structure S that plays *all* of the roles of spacetime in general relativity, at least under appropriately restricted conditions. General relativity has been so successful that it would be surprising if its roles for spacetime were vindicated only in part. One can reasonably hope to find some S that at least approximates the full relativistic role of spacetime. To do this, we can take the approach suggested by David Yates in this volume: Ramsify the full theory of general relativity, identifying spacetime with whatever plays the full role of spacetime in this theory, and then find a structure S that plays this full role under quantum gravity. On this approach, we need not settle which functional roles of spacetime are essential for spacetime in advance. We use something close to the full functional role to find spacetime in quantum gravity.

Still, what if it turns out (perhaps surprisingly) that under quantum gravity, no structure plays all aspects of the spacetime role? Suppose that a non-fundamental structure S plays some aspects of the standard spacetime role but not others. Then whether we count S as spacetime or not will depend on which functional analysis of spacetime we endorse. For example, S might count as spacetime on an operational analysis but not on an inertial analysis. In practice this may end up as a largely verbal dispute, which we can resolve by distinguishing multiple notions of spacetime. For example, we may distinguish inertial spacetime and operational spacetime, and make the case that the theory vindicates one but not the other.

Among specific functional analyses, it is not obvious that Knox's inertial spacetime functionalism is well suited to finding spacetime with quantum gravity theories. The reason is that Knox's analyses of the spacetime role use some spatiotemporal notions, for example in defining an inertial reference frame. But then, to find spacetime in quantum gravity by finding a structure S that plays the spacetime role, we will already need to have found some spatiotemporal properties within the framework. So we will at least need some further analyses of these spatiotemporal properties to support a reduction.

We can also use commonsense functionalism about space to help locate ordinary space and time under quantum gravity theories. Presumably there will be certain non-fundamental properties that tend to bring about ordinary spatial and temporal experience, at least in relevant contexts. On a phenomenal analysis, these can then be identified with ordinary space and time. One can also use a non-phenomenal analysis, finding non-fundamental properties that play the appropriate folk physics roles in motion and interaction. One can even combine elements of theoretical and commonsense approaches. For example, we could first make the case that in quantum gravity theories, certain non-fundamental properties play the key theoretical roles from general relativity or even from Newtonian physics (in relevant contexts). From prior analysis, we know it is plausible that whatever plays these theoretical roles also plays the commonsense roles. In this way, the intra-theoretical project of finding one theory's spacetime within a later theory's picture of the world can help with the Sellarsian project of finding the manifest image within the scientific image.[7]

References

Albert, D. (1996). Elementary quantum metaphysics. In J. T. Cushing, A. Fine, and S. Goldstein (Eds.), *Bohmian Mechanics and Quantum Theory: An Appraisal*, pp. 277–284. Dordrecht: Springer.

Albert, D. (2015). *After Physics*. Cambridge, MA: Harvard University Press.

Albert, D. (2019). Preliminary considerations on the emergence of space and time. In A. Cordero (Ed.), *Philosophers Look at Quantum Mechanics*, pp. 87–95. Cham: Springer.

Baker, D. J. (2019). On spacetime functionalism. http://philsci-archive.pitt.edu/15860.

Baron, S. (2019). The curious case of spacetime emergence. *Philosophical Studies 177*(8), 2207–2226.

Byrne, A. (2006). Color and the mind-body problem. *Dialectica 60*(2), 223–244.

Chalmers, D. J. (2005). The matrix as metaphysics. In C. Grau (Ed.), *Philosophers Explore the Matrix*, pp. 132–176. New York: Oxford University Press.

Chalmers, D. J. (2006). Perception and the fall from eden. In T. S. Gendler and J. Hawthorne (Eds.), *Perceptual Experience*, pp. 49–125. New York: Oxford University Press.

Chalmers, D. J. (2011). Verbal disputes. *Philosophical Review 120*(4), 515–566.

Chalmers, D. J. (2012). *Constructing the World*. Oxford: Oxford University Press.

Chalmers, D. J. (2017). The virtual and the real. *Disputatio 9*(46), 309–352.

Chalmers, D. J. (2018). Structuralism as a response to skepticism. *Journal of Philosophy 115*(12), 625–660.

Chalmers, D. J. (2019). Three puzzles about spatial experience. In *Blockheads! Essays on Ned Block's Philosophy of Mind and Consciousness*, pp. 109–137. Cambridge, MA: MIT Press.

Dees, M. K. (2015). *The fundamental structure of the world: Physical magnitudes, space and time, and the laws of nature*. Ph.D. thesis, Rutgers University-Graduate School-New Brunswick. https://rucore.libraries.rutgers.edu/rutgers-lib/47350/PDF/1.

Fredkin, E. (2000). Digital mechanics. Draft of December 2000.

[7] Sections 7.1 to 7.6 are an edited version of my third Jean Nicod lecture at the Ecole Normale Superieure in Paris on June 25, 2015. (Video at http://savoirs.ens.fr/expose.php?id=2167.) The Nicod lectures will eventually appear as a book from MIT Press. This material was first presented in talks to the ANU Research School of Physics and a Rochester conference on the metaphysics of physics in 2013, and later in talks at Arizona, Brown, Connecticut, Johns Hopkins, Oxford, and Rutgers. Thanks to audiences on all these occasions for their feedback. For comments, thanks also to Baptiste Le Bihan, Trevor Teitel, Christian Wüthrich, and David Yates.

Jackson, F. (1998). *From Metaphysics to Ethics: A Defence of Conceptual Analysis.* Oxford: Oxford University Press.

Knox, E. (2014). Spacetime structuralism or spacetime functionalism. Unpublished draft available at https://www.eleanorknox.com/uploads/3/9/8/2/39826525/structuralism_or_functionalism_draft_2. pdf.

Knox, E. (2019). Physical relativity from a functionalist perspective. *Studies in History and Philosophy of Science Part B: Studies in History and Philosophy of Modern Physics 67*, 118–124.

Lam, V. and C. Wüthrich (2018). Spacetime is as spacetime does. *Studies in History and Philosophy of Science Part B: Studies in History and Philosophy of Modern Physics 64*, 39–51.

Le Bihan, B. (2018). Space emergence in contemporary physics: Why we do not need fundamentality, layers of reality and emergence. *Disputatio 10*(49), 71–95.

Le Bihan, B. (2019). Spacetime emergence in quantum gravity: Functionalism and the hard problem. *Synthese*, 1–23. https://doi.org/10.1007/s11229-019-02449-6.

Lewis, D. K. (1994). Reduction of mind. In S. Guttenplan (Ed.), Oxford: *Companion to the Philosophy of Mind*, pp. 412–431. Blackwell.

Maudlin, T. W. (2007). Completeness, supervenience and ontology. *Journal of Physics A: Mathematical and Theoretical 40*(12), 3151.

Ney, A. (2015). Fundamental physical ontologies and the constraint of empirical coherence: a defense of wave function realism. *Synthese 192*(10), 3105–3124.

Read, J. and T. Menon (2019). The limitations of inertial frame spacetime functionalism. *Synthese*, 1–23. https://doi.org/10.1007/s11229-019-02299-2.

Rosenberg, G. (2004). *A Place for Consciousness: Probing the Deep Structure of the Natural World.* New York: Oxford University Press.

Seiberg, N. (2006). Emergent spacetime. arXiv preprint hep-th/0601234.

Sellars, W. S. (1962). Philosophy and the scientific image of man. In R. Colodny (Ed.), *Science, Perception, and Reality*, pp. 35–78. Ridgeview: Humanities Press.

Smith, B. (1996). *On the Origin of Objects*. Cambridge, MA: MIT Press.

Thompson, B. (2003). *The nature of phenomenal content*. Ph.D. thesis, The University of Arizona.

Thompson, B. (2010). The spatial content of experience. *Philosophy and Phenomenological Research 81*(1), 146–184.

Wilson, A. Explanations of and in time. In C. Wüthrich, B. Le Bihan, and N. Huggett (Eds.), *Philosophy Beyond Spacetime*. This volume.

Wolfram, S. (2003). A new kind of science. *Applied Mechanics Reviews 56*(2), B18–B19.

Yates, D. Thinking about spacetime. In C. Wüthrich, B. Le Bihan, and N. Huggett (Eds.), *Philosophy Beyond Spacetime*. This volume.

8
Explanations of and in Time

Alastair Wilson

Abstract

Various approaches to quantum gravity render spacetime an emergent phenomenon, with the existence and properties of spacetime depending on a non-spatiotemporal underlying reality. This chapter investigates the mode of dependence that is involved. I explain and defend my recent proposal to classify different kinds of dependencies in terms of the principles mediating the dependency, and apply this proposal to the emergence of spacetime. While philosophers have typically interpreted spacetime emergence as involving metaphysical grounding, I show how premises that are widely endorsed lead us to classify the emergence of spacetime in loop quantum gravity as causal in nature. I recommend spacetime functionalism as a resolution of this puzzle that vindicates the natural view of spacetime emergence as non-causal. I then explore a different approach to spacetime emergence in quantum gravity, the 'many-instant landscape' scenario described by Gomes, and show how it fits into the proposed framework for spacetime emergence.

8.1 Introduction

This volume explores conceptions of physical reality, suggested by theories of quantum gravity, in which spacetime as we know it is not fundamental. Theorists who wish to regard spacetime as real but non-fundamental typically speak of it as *emergent*. In this chapter I shall investigate how this emergence works.

The general notion of emergence is hugely contested, but we can extract two clear ingredients of the notion as it is applied within philosophy of science:

- *Dependence:* Emergent phenomena depend on that from which they emerge.
- *Novelty:* Emergent phenomena are novel relative to that from which they emerge.

Alastair Wilson, *Explanations of and in Time* In: *Philosophy Beyond Spacetime: Implications from Quantum Gravity.* Edited by: Christian Wüthrich, Baptiste Le Bihan, and Nick Huggett, Oxford University Press.
DOI: 10.1093/oso/9780198844143.003.0008

Proponents of 'strong emergence' require some kind of ontological or metaphysical novelty in the emergent phenomena; proponents of 'weak emergence' require only novel epistemic or explanatory features. But how exactly to understand the Novelty ingredient need not detain us here; instead I want to focus on the Dependence ingredient. How exactly does the spatiotemporal depend on the non-spatiotemporal? Or to put it another way, how exactly does the non-spatiotemporal give rise to the spatiotemporal?

Contemporary metaphysics acknowledges a variety of different forms of dependence. The most familiar form is causal dependence. It may seem obvious that emergent spacetime is not caused to exist. I think that, in the end, this apparently obvious thought can be upheld. But as we shall see, it turns out to be surprisingly difficult to square with mainstream assumptions about how metaphysics and physics interact.

If spacetime emergence is not causal, what is it? A natural thought is that spacetime emergence is an instance of an inter-level relationship as conceived in the classical hierarchical image of science, where different sciences deal with different levels of phenomena of varying fundamentality.[1] We then have a spatiotemporal level that depends on a non-spatiotemporal level. Of course, there are many ways of understanding talk of levels, and many authors reject such talk altogether (see e.g. Thomasson 2014). The classical approach is that of Oppenheim and Putnam (1958), who characterize levels in mereological terms. A more general approach that is currently popular is to theorize about levels in terms of grounding (Schaffer 2009; Fine 2012); facts at the microphysical level are said to ground facts at the macrophysical level.[2] Reduction, whether this is then cashed out in terms of derivability or in some other way, provides a further set of theoretical tools to approach inter-level relations.

In what follows, I will work with the notion of grounding, although not much hangs on this and I hope my arguments will be intelligible to those who prefer other frameworks for understanding inter-level relations in science. Readers may regard my talk of grounding in this chapter as a placeholder for whatever non-causal relations connect levels of nature. For present purposes I am simply concerned to make conceptual space for the possibility that spacetime emerges non-causally.

We can now pose the question: is dependence of the spatiotemporal on the non-spatiotemporal an instance of causation or of grounding? Answering this question requires saying something general about how to distinguish instances of causation from instances of grounding, and on this point the literature is surprisingly

[1] In any realistic picture, these levels are likely to be vaguely characterized and only partially ordered.

[2] The grounding approach may subsume the mereological approach, if parts ground their wholes (Cameron 2014). I will not address priority monism (Schaffer 2010) which inverts the relationship between levels.

thin. The distinction between grounding and causation is typically introduced by reference to paradigm cases and then presupposed without further explanation (Wilson 2018, 2020). There are clear cases of causation: collisions between billiard balls, initiations of chemical reactions, nutritional factors influencing disease. There are clear cases of grounding: the dependency of the pressure of a gas on the motion of its molecular constituents, the dependency of a set on its members, and the dependency of the permissibility of an action on the intentions of an agent. But is the dependency of spacetime on non-spatiotemporal reality more like the first kind of case or more like the second kind of case? The question is intractable as it stands; we need a general criterion for telling causation apart from grounding.

In the next section, I will criticize a temporal criterion for distinguishing grounding from causation and identify an alternative criterion—the *mediation criterion*—as better suited to understanding cases of dependent spacetime. In section 8.3 I argue that the application of the mediation criterion is liable to classify emergent spacetime in loop quantum gravity (LQG) as caused rather than as grounded, but in section 8.4 I show how functionalism about spacetime provides a way of avoiding this unexpected, and potentially unwelcome, conclusion. Section 8.5 applies the framework of sections 8.2 to 8.4 to a case study of an emergent spacetime theory, the many-instant landscape view developed by Henrique Gomes. Section 8.6 is a conclusion.

8.2 Demarcation Criteria

In recent work I have explored a number of candidate criteria for distinguishing causation from grounding, arguing that the most obvious criteria fail (Wilson 2020). The most widespread (although usually implicit) of these criteria is what I call the *temporal criterion*. According to the temporal criterion, causation is diachronic—a cause and its effect occur at different times—while grounding is synchronic—a grounding fact and the fact that it grounds are facts about the same time.

The temporal criterion faces a number of difficulties. Elsewhere I have criticized it by appeal to apparent counterexamples in both directions—cases of simultaneous causation and of diachronic grounding[3]—but in the present context we face a problem even in applying the criterion. In a case of dependence of a spatiotemporal non-fundamental reality on a non-spatiotemporal fundamental reality, the facts involved do not either occur at the same time *or* occur at different times. The non-spatiotemporal fundamental reality, as the label suggests, doesn't occur in

[3] Sample cases: Newtonian forces simultaneously cause instantaneous accelerations, and Obama *is* a former president because he *was* a president. See Wilson (2020) for further discussion.

time at all. The spatiotemporal non-fundamental reality doesn't occur in time either: if anything, time occurs in it. In cases of emergent spacetime it seems that the dependency involved in the emergence is neither diachronic nor synchronic, but rather timeless or *achronic*. The facts related by the dependency stand in no temporal relations at all to each other. For this reason, emergent spacetime hypotheses present a distinctive difficulty for the temporal criterion.

A potential response to this difficulty is to recast the temporal criterion into a *modified temporal criterion*: causation is diachronic while grounding is non-diachronic. Then both synchronic and achronic cases are classified together, and jointly contrasted with diachronic cases. While this avoids the immediate difficulty, it may well seem suspiciously ad hoc. Nor does the response resolve all our problems: in recent work (Wilson 2020) I describe potential cases of synchronic causation (e.g. $F = ma$ in classical mechanics) and of diachronic grounding (e.g. cross-time consequences of externalist individuation criteria), and these cases cause trouble even for the modified temporal criterion. A further reason to worry about the modified criterion is that it rules out the conceptual possibility that the whole universe should be caused to exist, or that it should be intervened upon by an atemporal deity. Most importantly, in the present context the retreat to the modified temporal criterion is dialectically unsatisfying: it amounts to a bare stipulation that emergent spacetime and other forms of achronic dependence are grounding rather than causation. A principled basis for this classification is what we are after.

An alternative potential criterion that may occur to some readers is what I have called the distinctness criterion (Wilson 2020). According to the distinctness criterion, causation is a dependency relation between fully distinct events, while grounding is a relation between not fully distinct events. This criterion is in effect used by Baron and Miller (2015) as at least a necessary condition[4] for causation in their discussion of causation in a universe with spatial but no temporal structure; they are drawing here on Lewis (1973). Baron and Miller argue that if two fully distinct events stand in a nomic dependency relation, then dependencies between them can count as causal even in the absence of any temporal relations. I have argued against the distinctness criterion elsewhere (Wilson 2020); for now, I note that (as with the temporal criterion) the emergent spacetime scenario presents a distinctive difficulty for the distinctness criterion. The distinctness criterion as employed by Baron and Miller and by Lewis explicitly concerns dependence between events. Baron and Miller undertake to argue that we can identify distinct events even in the timeless scenario. However, when we

[4] Baron and Miller do not directly address whether there can be grounding relations between distinct events, but the tenor of their discussion suggests they assume not; this would amount to full endorsement of the distinctness criterion.

are dealing with the emergence of spacetime from a wholly non-spatiotemporal fundamental level (as in LQG), there may be no things identifiable as distinct events at the level of the fundamental physics at all; or at least, the basic physical elements of the theory will lack much of the structure we typically associate with events. So emergent spacetime hypotheses present special problems both for defenders of the temporal criterion and for defenders of the distinctness criterion as general accounts of the difference between grounding and causation.

It is time to turn to the approach that I favour for distinguishing between causation and grounding: the *mediation criterion*, which I have defended elsewhere (Wilson 2018, 2020). The mediation criterion makes use of the notion of a *mediating principle:* a general principle that mediates the connection between a particular dependent fact and the particular fact it depends on. The motivating picture here is that dependencies are not scattered piecemeal or at random about reality, but rather hold in a systematic way that is captured by a limited set of general principles. Whenever there is a dependency between two facts, there is some general principle in the vicinity linking facts like the former with facts like the latter. This does not amount to an endorsement of a general principle of sufficient reason: I am content to allow that many facts may have no explanation at all and that even dependency facts may only have partial explanations. The assumption we need is just that the holding of individual instances of dependency can always be at least partially accounted for in general terms, and I regard that as extremely plausible.

According to the mediation criterion, causal dependencies are dependencies that are mediated by some law of nature, while grounding dependencies are dependencies that are not mediated by any law of nature. Typically grounding dependencies are instead mediated by what I will call *constitutive principles:*[5] principles that tell us what it is to be a particular thing or kind of thing. (In both cases there will also be a role for background conditions of various kinds; set these aside for the time being.) The mediation criterion transforms questions about the nature of a dependency connection into questions about the nature of the general principles that mediate it. The idea behind it is that we can use the grasp we have on the notion of a law of nature to cast light on the distinction between causation and grounding.

Examples will help to illustrate the application of the mediation criterion. A paradigm particular case of causation would be a collision between two specific

[5] There is an interesting question here about whether logical and mathematical dependencies are mediated by constitutive principles, or by some other kind of principle. (They evidently are not mediated by any laws of nature, so they will not count as causal, however we resolve their status.) One possible account of the difference between logical and metaphysical dependence would be that logical dependencies are mediated by inference rules, whereas metaphysical dependencies are mediated by factual principles. I explore these matters elsewhere (Wilson MS).

billiard-balls at some specific time, as described by classical mechanics. It is no coincidence that the resultant motion of ball *B* depends on the initial motion of ball *A*: energy and momentum are conserved in classical mechanics, and so the laws of classical mechanics determine that if the initial motion of ball *A* is varied, then the resultant motion of ball *B* varies along with it. Here is a clear case of the dependency between two variables being explained by general principles: the dependency between the motions of the billiard-balls is being explained by the operation of the laws of classical mechanics. Thus the mediation criterion classifies this case as an instance of causal dependence. In contrast, a paradigm case of grounding is the dependence of the existence of the singleton set containing Socrates on the existence of Socrates himself. It is of course no mere coincidence that the existence of Socrates' singleton depends on the existence of Socrates: the principles of impure set theory dictate that all and only existing entities have a singleton set. Accordingly, if Socrates were to be somehow excised from reality, then this singleton set would disappear along with him. Thus the principles of impure set theory explain why there's a connection between the philosopher and the set. No law of nature is involved in mediating the dependence, and so the mediation criterion counts this as a case of grounding.

So far, so good. Of course, once we move beyond paradigm cases, matters are not so clear-cut. In particular, moving to unfamiliar physical contexts tests our grasp of the distinction between laws and non-laws. The best way to find out whether our grasp of this distinction is adequate for its assigned role is to investigate its application in some concrete cases. In the next section I turn to the application of the mediation criterion to the emergence of spacetime in the specific context of the LQG family of quantum theories of gravity. Further applications with which to contrast this case would certainly be valuable—the case of renormalization group explanations is a particularly interesting application to consider—but these will have to await further work. This chapter focuses on emergent spacetime, and it relies on whatever grasp of the law/constitutive principle distinction we can muster in that context.

8.3 The Mediation Criterion in Loop Quantum Gravity

A major consideration that favours the approach to classifying dependencies that is built around the mediation criterion is that the criterion extends seamlessly to atemporal forms of dependency: unlike the temporal criterion, it does not presuppose a background temporal structure in order to classify dependencies as causal or as grounding. There is room, therefore, to apply the mediation criterion to the emergence of spacetime in quantum gravity. I will consider in particular the application of the mediation criterion to the emergence of spacetime in LQG.

The reason for this choice is that I take LQG to be the most well-developed approach in which spacetime is fully emergent: on typical interpretations, LQG assumes less fundamental spatiotemporal structure than its main rival, string theory.[6] LQG accordingly presents the most radical conceptual break with the classical conception of spacetime as fundamental constituents of reality. In section 8.5 I will turn to a different approach to quantum gravity, Henrique Gomes' 'many-instant landscape' in which time is emergent but space is not, so that the different applications of the mediation criterion may be contrasted.

As I understand LQG,[7] the aim is to describe a fundamental non-spatiotemporal reality, the physical behaviour of which gives rise to 'effective' spatiotemporal structure at a non-fundamental level. Accordingly, what is posited is the dependence of the spatiotemporal aspects of the world on some non-spatiotemporal aspects of the world. Applying the mediation criterion, we ask what principles mediate the connection between the non-spatiotemporal facts and the spatiotemporal facts. Such principles would explain why, given the non-spatiotemporal facts, we obtain the spatiotemporal facts. A first-pass schematic answer extractable from LQG is the following:

> *LQGP*: If there is a suitable superposition of spin foams, then spacetime exists and can be approximately represented by a model of General Relativity.

It is up to the proponents of LQG to fill out the details of this schema. Wüthrich (2017) describes one way in which these details may be spelled out, drawing on work by Butterfield and Isham (1999, 2001) and by Ashtekar, Rovelli, and Smolin (1992). For LQG to be viable, its proponents will need to be able to specify in general terms what features are required for a superposition of spin foams to give rise to familiar four-dimensional spacetime as approximated by particular models of General Relativity—as well, of course, as being able to show that such features are present in well-behaved and well-understood models of LQG. Assuming LQG proponents succeed in identifying such features of superpositions of spin foams, they will have identified some candidate physical principles connecting the fundamental reality of spin foams with the emergent reality of spacetime.

What, then, is the nature according to LQG of the principles schematized by LQGP?—Are they laws of nature, or are they constitutive principles? It would be wrong to assume that because these principles are posited as part of a candidate

[6] This question is complicated by the interpretation of dualities in string theory; considerations of duality have suggested to some that spacetime is (in my sense) fully emergent in string theory also. See Huggett (2017) and Le Bihan and Read (2018) for discussion.

[7] I will not need to get into any technical details of LQG. The version of the theory with which I am most familiar is that of Rovelli (2004), but I think that the aspects of the theory that I will discuss are common to most approaches.

physical theory they must be laws of nature. Theories that have historically been central to physics, such as the atomic theory of matter and the wave theory of light, involve constitutive principles as well as laws of nature. The principles concerning the interactions of atoms are laws of nature and mediate causal connections, but the principles concerning the dependency of bulk properties of macroscopic matter on the collective behaviour of the matter's atomic constituents are constitutive principles and mediate grounding connections. The principles concerning the propagation of waves are laws of nature and mediate causal connections, but the principles concerning the dependency of light colour on wavelength are constitutive principles and mediate grounding connections.

It is a natural thought that the emergence of spacetime is a form of grounding, not causation. The mediation criterion accordingly predicts that it should be a natural thought that the principles concerning the emergence of spacetime should be constitutive principles, and not laws of nature. And this prediction is borne out; it is a natural thought that instances of the LQGP schema are principles about the underlying nature of spacetime. Detailed interpretive work on LQG, of the sort referenced above, is then conceived as telling us about how spacetime is made up or constituted according to LQG. This initially plausible-seeming picture, however, is immediately threatened by a clash with the conjunction of two widely held assumptions within metaphysics and philosophy of physics respectively: that constitutive principles are metaphysically non-contingent and that Newtonian spacetime is metaphysically possible.

The problem, in short, is that the following four claims cannot all be true together:

1. Spacetime is constituted by a superposition of spin foams.
2. Constitution of spacetime is metaphysically non-contingent.
3. Newtonian spacetime is metaphysically possible.
4. Newtonian spacetime is not constituted by a superposition of spin foams.

If the constitution of things is in general non-contingent (as metaphysicians almost uniformly suppose it is), then there is only one possible way in which spacetime can be constituted, and so either emergent spacetime as in LQG or fundamental spacetime as in Newtonian mechanics must be impossible.[8]

There are routes out of this puzzle which involve giving up widely endorsed metaphysical assumptions. In particular, we might give up 2 and reject the

[8] Note that even if we regard Newtonian mechanics as inconceivable on the grounds (for example) that it does not allow for stable matter, the present problem remains unresolved since it concerns fundamental spacetime rather than any particular theory in such spacetime. For theoretical pathologies to resolve the present problem, there would need to be something irredeemably wrong with every theory that posits a fundamental spacetime; this seems improbable.

Kripkean principle of the necessity of constitution. Or, we might give up 3 and reject the inference from conceivability to possibility.[9] These moves may seem tempting, and especially so to those with no sympathy for contemporary metaphysics. So, would either move be justifiable as naturalistic metaphysics? I think not. In the present dialectical context, such moves would be out of proportion; whether there is or is not good reason to reject 2 or 3, such reason is unlikely to turn on anything specific to spacetime. After all, LQG is a theory about spacetime, not a theory about constitution or about modality. Without some independent argument against 2 or 3 and without some explanation of why they have seemed plausible despite being false, the moves just look ad hoc. Assuming, then, that we retain 2 and 3, we can break the argument down as follows:

1. Spacetime emergence is mediated by LQGP.
2. Constitution of spacetime is metaphysically non-contingent.
3. Newtonian spacetime is metaphysically possible.
4. If spacetime were Newtonian, LQGP would be false.
5. LQGP is not a constitutive principle.
6. If LQGP is not a constitutive principle, it is a law of nature.
7. Spacetime emergence in LQG is causation, not grounding.

Could we live with this conclusion? Perhaps; but it would at least be somewhat surprising, and so it is worth exploring potential ways of avoiding the result.

8.4 Spacetime Functionalism

The best way of avoiding the conclusion that spacetime emergence is causal, without being forced into ad hoc metaphysics, seems to me to be to adopt functionalism about spacetime. If to be spacetime is to occupy a certain functional role within physical reality, and if that functional role can be occupied by entities of varying underlying constitution—that is, if spacetime can be multiply realized—then our conclusion 7 can be avoided and spacetime emergence can once more be regarded as grounding rather than as causation.

Functionalism about X is the view that to be X is to have a certain characteristic behaviour or function. Functionalists distinguish between *roles* and *realizers*: the role is the function itself, and the realizer is the thing that functions. Functionalism

[9] I should note that I have argued that 3 is false on completely independent grounds (Wilson 2013). The present chapter adopts more orthodox assumptions about modality for the sake of the argument.

about spacetime, then, is the view that to be spacetime is to have characteristic spatiotemporal behaviour. Functionalism about spacetime blocks the argument for spacetime emergence as causal. Recall the first two premises:

1. Spacetime emergence is mediated by LQGP.
2. Constitution of spacetime is metaphysically non-contingent.

In the functionalist picture, 1 and 2 cannot be true together on a uniform reading of 'spacetime'. If we focus on the realizer of the spacetime role, then 1 is true (since the emergence of that actual realizer is mediated by LQGP) but 2 is false: it is contingent which realizer plays the spacetime role. If we focus on the spacetime role itself, then 2 is true (since the nature of the spacetime role is non-contingent) but 1 is false: the connection between the physical facts and the existence of spacetime is now mediated by a *functional constitutive principle*, and this functional principle (unlike LQGP) does not make reference to any specific realizer.

What is a functional constitutive principle? In my usage, it is a principle that specifies the functional role in question by connecting the performance of that function with the existence of the functionally constituted phenomenon. In computational functionalism about the mind, the relevant principle might be that to think is to perform certain computations. In functionalism about life, the principle might be that to be alive is to maintain homeostasis, to metabolize, etc. In a functionalism about seats, perhaps the relevant principle might be that to be a seat is to be something that can be sat upon. In each case, the functional constitutive principle in question says that what it is for some functionally constituted fact to obtain is for some particular role to be realized.

It has been suggested to me by Carl Hoefer and by Eddy Keming Chen that instead of functional constitutive principles we might appeal at this point to a determinable/determinate structure: *being a spacetime* would be a determinable property and *being a Newtonian spacetime* would be a determinate of the spacetime determinable, with *being an emergent spacetime in LQG* a distinct determinate of the same determinable. My reasons for preferring to take the functionalist route are not specific to the emergent spacetime context; I take it to be a general scientific methodological principle that where possible it is desirable to account for determinable/determinate structures in terms of functional resemblance across the different determinates. Here I would echo Dennett (2001, 233): "Functionalism in this broadest sense is so ubiquitous in science that it is tantamount to a reigning presumption of all of science". However, I don't take this to be decisive; while it may be unfamiliar to regard *being a spacetime* as a determinable property, doing so remains a live option.

In the spacetime functionalism context, the million-dollar question is: what is the functional role of spacetime? The most influential proposal in the present literature is that of Eleanor Knox:[10]

> a structure will play the spacetime role in our theories just in case it describes the structure of the inertial frames, and the coordinate systems associated with these. (Knox 2014)

Knox is in effect proposing that to be spacetime is to be whatever element of reality it is that determines inertial motion of matter. It is easy to see how this would be a different element of reality depending on whether Newtonian mechanics or LQG were correct. In LQG (let us assume for the sake of argument) certain non-fundamental structural features of superpositions of spin foams determine the inertial motions of matter. In Newtonian mechanics, fundamental absolute space determines the inertial motions of matter. Hence spacetime is differently realized according to the two different theories, yet retains a single functional constitution across both cases.

In my view the ability to seamlessly handle a variety of hypotheses about spacetime, including theories that make spacetime fundamental and theories that make it emergent, is the most powerful consideration in favour of spacetime functionalism. Functionalism provides the best way to make clear sense of the idea that theories with such radically different conceptions of the underlying nature of spacetime are nevertheless theories about one and the same thing. This consideration was already forceful enough when comparing different theories of non-quantum spacetime, but the radical variety of live theoretical possibilities within the quantum gravity literature makes spacetime functionalism seem close to inevitable if we are to compare the various different approaches.[11]

Spacetime functionalism provides one clear way to regard spacetime as grounded rather than caused. There are other ways: for example, the hypothesis of a determinable-determinate structure of possible spacetimes, or a more-or-less-ad-hoc rejection of premises 2 and 3. But to adopt any such approach is still merely to indicate the shape of an answer to the question of how spacetime emerges in quantum gravity. Once we have found a way to regard spacetime as grounded, we still need to understand the details of what it is grounded in and which functional constitutive principles mediate the grounding dependency.

Theories of quantum gravity that incorporate emergent spacetime posit a non-spatiotemporal fundamental reality that underlies spatiotemporal reality.

[10] Knox is here building on ideas of Brown (2005). See also Lam and Wüthrich (2018), who offer an alternative characterization of the spacetime role which they argue to be more general than Knox's.

[11] This point is emphasized by Lam and Wüthrich (2018).

The non-spatiotemporal reality is intended to be fully physical in nature, even though it lacks spatiotemporal properties; so it is plausible to think it continues to be governed by physical laws of nature relating its different elements, whatever exactly those elements may be. According to the mediation criterion, any dependencies mediated by laws of nature are causal dependencies. This line of thought suggests that non-trivial physical theories that describe a non-spatiotemporal fundamental reality will typically still involve extensive causal dependencies between different elements of that reality. In such a scenario, then, we would have a general answer to our question about the grounds of spacetime: spacetime is grounded in causal processes at the fundamental level.

It is frequently assumed that there can be no causation without space and time.[12] However, it seems to me that this assumption stems from the mistaken assimilations of causation with diachronic dependence and of grounding with synchronic dependence. Once we recognize the potential for the causation/grounding distinction to come apart from the diachronic/synchronic (or diachronic/non-diachronic) distinction, we see that there is the potential to draw a principled line between non-diachronic dependencies that are cases of causation and non-diachronic dependencies that are cases of grounding. Non-diachronic causal dependencies will be mediated by laws of nature whose essential role is not that of evolving a physical system in time, but rather of determining (timelessly) the values of certain 'output' physical variables in terms of 'input' physical variables. Such generalized laws may be unfamiliar, but I see nothing impossible in them.[13]

An argument could even be made at this point that a fundamental physical theory *must* posit at least some causal dependencies between distinct physical variables: perhaps a theory which consists only of constitutive principles, without including any laws of nature, would not count as a physical theory at all but rather a metaphysical theory. While I think this argument has some merit, I will not explore it further here. What I want to focus on instead is a specific way of implementing the general thought that spacetime is grounded in non-spatiotemporal physics. One illustrative proposal concerning the functional grounds of spacetime in quantum gravity, which features in recent work by Henrique Gomes, is considered in the next section. In Gomes' proposal (Gomes 2017), space is fundamental but time is not, and temporal orderings are reconstructed by appeal to dynamical properties of particular subsystems that are capable of functioning as (relational)

[12] Baron and Miller (2014, 2015), who have argued for the possibility of causation in a timeless setting, are a notable exception within the metaphysics literature. Their arguments are complementary to my arguments in this section, although they are working with the distinctness criterion rather than with the mediation criterion. In the quantum gravity literature, causal set theory (e.g. Dowker 2013) may be seen as an alternative implementation of the idea that causal relations are more fundamental than spatiotemporal relations and may exist in the absence of the latter.

[13] Nor do Baron and Miller (2015).

clocks. The proposal fits neatly into the general model for understanding functional emergence of spacetime that I have been defending in this chapter.

8.5 Grounding the Spatiotemporal in the Many-Instant Landscape

Gomes (2017) models possible quantum histories in a 'timeless' state space of *spatial field configurations*. He calls this the 'many-instant landscape'. To calculate transition probabilities, Gomes uses path integrals along paths from an 'initial' boundary condition, through a 'record' state, to a 'resultant' state. These paths cannot in general be interpreted as temporal histories. Some segments of these paths allow for stable clock behaviour, giving rise to an emergent temporal domain, but not all path segments do. States in the 'deep quantum' domain do not stand in any temporal relations to any other states.

Gomes himself takes causality to be conceptually linked closely to time, and in particular to the relativistic notion of timelike intervals, and accordingly he restricts the term 'causal' to apply only within the emergent temporal domain. For reasons explained earlier in the chapter, I prefer to use 'causal' more broadly; in my preferred usage from sections 8.1 to 8.4, all of the physically permitted[14] histories in Gomes' picture are causal histories, since there are law-of-nature-mediated explanatory connections between the states along these histories, even though only some of the histories count as temporal histories. However, to avoid confusion in the remainder of this section I shall adopt a terminology that does not clash with Gomes' own usage and contrast temporal histories with *dynamical histories*.[15] In these new terms, what the many-instant landscape picture aims to do is to account for the temporal ordering of states in terms of a more fundamental underlying dynamical ordering that makes essential appeal to a distinctive boundary condition.

Gomes takes inspiration from Boltzmannian approaches to the thermodynamic temporal asymmetry. In Boltzmannian approaches, a very special low-entropy boundary condition is allied with dynamical laws and a probability measure over states to provide an explanation of the observed temporal asymmetry of entropy. The hypothesis of this special boundary condition is known as the *Past Hypothesis*:[16]

[14] Gomes' framework also allows us to characterize generic histories, subject to no equation-of-motion constraints; such histories do not in general display any law-like behaviour and I would not be inclined to regard them as physically possible histories, although different approaches to the metaphysics of these histories are no doubt possible.

[15] The word 'dynamical' has unfortunate connotations of change, and correspondingly of temporal succession. The intended use of the term here is opposed to 'kinematical'; it aims to indicate substantive constraints imposed by physics on the world, without having any implications for time or change.

[16] David Albert's *Time and Chance* (2000) is the source of this term, and the *locus classicus*.

- *Past Hypothesis*: A low-entropy boundary condition obtains at one temporal end of the universe.

We label the low-entropy end of the universe the 'initial state'; however, as Price (1997) emphasizes, we do not have any grip on which end of the universe is initial and which is final that is independent of our physical description of the universe. Hence, in using the term 'initial state' we should be careful not to attribute any intrinsic direction to time. We have an asymmetry *in* time, not an asymmetry *of* time.

Gomes retains the general shape of the Boltzmannian account of temporal asymmetry, but extends it beyond domains where temporal orderings are well defined, replacing the Past Hypothesis with a boundary condition at the most homogeneous point of the underlying configuration space:

- *Dynamical Past Hypothesis*: A maximally homogeneous spatial field configuration φ^* obtains, and all dynamically permitted histories pass through it.

We now think in terms of a dynamically initial state instead of a temporally initial state, but as previously this should not be taken to imply any independent grip on the dynamical orientation of the universe. We have an asymmetry *in* the dynamical order rather than an asymmetry *of* the dynamical order.

There are clear affinities between the Past Hypothesis and the Dynamical Past Hypothesis. Each of them picks out a particular state as physically privileged, and as having a foundational role with respect to the calculation of transition probabilities between states that makes it indispensable for our physical understanding of the macroscopic asymmetry of time and causation. The key conceptual point, for present purposes, is that the Dynamical Past Hypothesis generalizes the Past Hypothesis by dropping the presupposition of a fundamental temporal ordering of states.

The resulting picture is one in which the temporal ordering of states is emergent rather than fundamental. What is fundamental is the dynamical ordering of states, and the associated transition probabilities. These fundamental physical elements then ground the existence of spacetime according to an appropriate functional constitutive principle, the details of which Gomes leaves open but describes in general terms as follows:

> space-times do not exist a priori, but require a relational construction of duration—or definitions of clocks—on classical solutions. i.e. in general one must define clocks and an experienced duration through the classical evolution of relational observables... (Gomes 2017, 5)

Where emergent time is well defined, the temporal ordering will align with the dynamical ordering. But the dynamical ordering is defined even in the deep quantum epoch, dynamically prior to the emergence of time. There need not be (and plausibly will not be) any sharp differences between these temporal and non-temporal domains; whether emergent time is well defined over a given sequence of instances may be a vague and messy business. No worries: all of the vagueness is restricted to the derivative level. The dynamical ordering remains well defined, and the fundamental level remains precise.

8.6 Conclusion

I have argued that emergent spacetime raises a distinctive challenge to standard views of how grounding and causation differ. Approaches that presuppose that space and time are fundamental aspects of reality need to be made more flexible in order to make coherent sense of theories of quantum gravity that posit a non-spatiotemporal fundamental reality.

The mediation criterion is one suitable generalized criterion for distinguishing grounding from causation. The criterion permits us to count spacetime emergence as grounding either by appeal to functionalism about spacetime or by appeal to a revisionary view of the nature of spacetime as non-contingent. The criterion also allows for fundamental physical reality to have pervasive causal structure even in the absence of pervasive temporal structure. LQG and Gomes' many-instant landscape provide two very different potential implementations of emergent spacetime, but each can be understood through application of the mediation criterion.[17]

References

Albert, D. Z. (2000). *Time and Chance*. Cambridge, MA: Harvard University Press.

Ashtekar, A., C. Rovelli, and L. Smolin (1992). Weaving a classical metric with quantum threads. *Physical Review Letters 69*, 237–240.

Baron, S. and K. Miller (2014). Causation in a timeless world. *Synthese 191*(12), 2867–2886.

[17] I am grateful for helpful discussions and comments on previous drafts to Baptiste Le Bihan, Richard Healey, Carl Hoefer, Nick Huggett, Eddy Keming Chen, Eleanor Knox, Katie Robertson, Kian Salimkhani, Antonio Vassallo, Chris Wüthrich, audiences in Chicago, Geneva, Bonn, and San Diego, and an anonymous referee for OUP. This work forms part of the project A Framework for Metaphysical Explanation in Physics (FraMEPhys), which received funding from the European Research Council (ERC) under the European Union's Horizon 2020 research and innovation programme (grant agreement no. 757295). Funding was also provided by the Australian Research Council (grant agreement no. DP180100105).

Baron, S. and K. Miller (2015). Causation sans time. *American Philosophical Quarterly 52*(1), 27–40.

Brown, H. R. (2005). *Physical Relativity: Space-Time Structure from a Dynamical Perspective*. Oxford: Oxford University Press.

Butterfield, J. and C. Isham (1999). On the emergence of time in quantum gravity. In J. Butterfield (Ed.), *The Arguments of Time*, pp. 111–168. Oxford: Oxford University Press.

Butterfield, J. and C. Isham (2001). Spacetime and the philosophical challenge of quantum gravity. In C. Callender and N. Huggett (Eds.), *Physics Meets Philosophy at the Planck Scale*, pp. 33–89. Cambridge: Cambridge University Press.

Cameron, R. P. (2014). Parts generate the whole, but they are not identical to it. In A. J. Cotnoir and D. L. M. Baxter (Eds.), *Composition as Identity*, pp. 90–110. Oxford: Oxford University Press.

Dennett, D. (2001). Are we explaining consciousness yet? *Cognition 79*(1–2), 221–237.

Dowker, F. (2013). Introduction to causal sets and their phenomenology. *General Relativity and Gravitation 45*(9), 1651–1667.

Fine, K. (2012). Guide to ground. In F. Correia and B. Schneider (Eds.), *Metaphysical Grounding: Understanding the Structure of Reality*, pp. 37–80. Cambridge: Cambridge University Press.

Gomes, H. (2017). Quantum gravity in timeless configuration space. *Classical and Quantum Gravity 34*(23), 235004.

Huggett, N. (2017). Target space ≠ space. *Studies in History and Philosophy of Modern Physics 59*, 81–88.

Knox, E. (2014). Spacetime structuralism or spacetime functionalism? Manuscript online at www.eleanorknox.com.

Lam, V. and C. Wüthrich (2018). Spacetime is as spacetime does. *Studies in History and Philosophy of Modern Physics 64*, 39–51.

Le Bihan, B. and J. Read (2018). Duality and ontology. *Philosophy Compass 13*, e12555.

Lewis, D. (1973). Causation. *Journal of Philosophy 70*(17), 556–567.

Oppenheim, P. and H. Putnam (1958). Unity of science as a working hypothesis. In H. Feigl, M. Scriven, and G. Maxwell (Eds.), *Concepts, Theories, and the Mind-Body Problem*, pp. 3–36. Minneapolis, MN: University of Minnesota Press.

Price, H. (1997). *Time's Arrow and Archimedes' Point: New Directions for the Physics of Time*. New York: Oxford University Press.

Rovelli, C. (2004). *Quantum Gravity*. Cambridge: Cambridge University Press.

Schaffer, J. (2009). On what grounds what. In D. Chalmers, D. Manley, and R. Wasserman (Eds.), *Metametaphysics: New Essays on the Foundations of Ontology*. New York: Oxford University Press.

Schaffer, J. (2010). Monism: The priority of the whole. *Philosophical Review 119*(1), 31–76.

Thomasson, A. L. (2014). It's a jumble out there: How talk of levels leads us astray. *American Philosophical Quarterly 51*(4), 285–296.

Wilson, A. (2013). Schaffer on laws of nature. *Philosophical Studies 164*(3), 653–667.

Wilson, A. (2018). Metaphysical causation. *Noûs 52*(4), 723–751.

Wilson, A. (2020). Classifying dependencies. In G. Darby, D. Glick, and A. Marmadoro (Eds.), *The Foundation of Reality: Fundamentality, Space, and Time*. Oxford: Oxford University Press.

Wilson, A. (MS). The gaps between the grounds. Available on request.

Wüthrich, C. (2017). Raiders of the lost spacetime. In D. Lehmkuhl, G. Schiemann, and E. Scholz (Eds.), *Towards a Theory of Spacetime Theories*, pp. 297–335. Basel: Birkhäuser.

9

Do You *See* Space? How to Recover the Visible and Tangible Reality of Space (Without Space)

Jenann Ismael

Abstract

In the search for a theory of quantum gravity, there are strong theoretical pressures that have pushed in the direction of theories in which space (or spacetime) is not present at the fundamental level. The task of recovering the appearances is especially pressing in such theories. This chapter looks at the cognitive processes that produce spatial experience to better understand the empirical constraints on such theories. There is no question that we have immediate awareness of the visible and tangible reality of space, but what that awareness amounts to, and whether it supports the requirement that space has to be recovered as concrete external structure, is not something that has received enough attention. This chapter fills that gap.

> If one asks what . . . is characteristic of the world of ideas of physics, one is first of all struck by the following: the concepts of physics relate to a real outside world, that is, ideas are established relating to things such as bodies, fields, etc., which claim 'real existence' that is independent of the perceiving subject-ideas which on the other hand, have been brought into as secure a relationship as possible with the sense-data. It is further characteristic of these physical objects that they are thought of as arranged in a spacetime continuum.
>
> (Einstein, Letter to Max Born, 5 April 1948 (section II, p. 170))

9.1 Introduction

When you open your eyes and look around, you see an evolving spatially structured world of objects that can be heard, felt, smelled, approached, and grasped.

Jenann Ismael, *Do You* See *Space? How to Recover the Visible and Tangible Reality of Space (Without Space)* In: *Philosophy Beyond Spacetime: Implications from Quantum Gravity.* Edited by: Christian Wüthrich, Baptiste Le Bihan, and Nick Huggett, Oxford University Press. © Jenann Ismael 2021. DOI: 10.1093/oso/9780198844143.003.0009

There are strong theoretical pressures coming from the search for a theory of quantum gravity to move to theories in which spacetime is not present at the fundamental level.[1] Recovering the appearances in such a theory is non-trivial, and has been largely presumed to have to take the form of showing how space (or spatiotemporal structures) emerges (at least approximately, in the appropriate limit) as a concrete structure, prior to the introduction of agents.[2] Very little explicit attention has been given in this setting to spatial perception. My chapter tries to fill that gap.

The first part looks at how the brain coordinates visual, tactual, and sensorimotor information to stabilize a conception of space as the common arena of both perception and action, and looks at what it means to *see space*, or to have immediate experience of the world as spatially structured. The second part draws some lessons for attempts to recover space and time in research in the foundations of quantum gravity.

The point of the chapter is to begin to bring the process that produces spatial experience into view so that we can better understand what the empirical constraints are for a theory in which spacetime does not appear at the fundamental level. Readers who want to cut to the lessons for quantum gravity research might start with the second part and work backwards.

9.1.1 Quantum Gravity

The problem of quantum gravity is the problem of obtaining a complete quantum description of gravitational phenomena. Gravity has been inextricably linked to the geometry of spacetime by General Relativity, and any quantum theory of gravity has to preserve that link. There are two main classes of approach: one that results from quantizing a classical theory of gravity[3] (i.e. General Relativity), and one in which spacetime and gravity are both emergent from something else.[4] The main problem for both programs is the empirical challenge of recovering classical space(-time) in the appropriate limit. This is a straightforward matter of empirical accuracy, since the world is manifestly spatially structured.[5]

[1] Those pressures are in evidence throughout this volume but see in addition the references in note 5.

[2] In practice, it is not actually easy to say what it is for spacetime to emerge because it is not clear which features of spacetime as it appears in classical physics we want to regard as necessary for a structure being called 'spacetime'. But it is presumed that it is at least essential that the full emergence of a classical spacetime does not require the introduction of agents. It is perhaps worth adding that we remain neutral on the question of whether space is discrete or continuous, as well as on the question of whether space is a substance or a network of relations.

[3] See Teitelboim (1982), Kiefer and Weber (2005), and Oriti (2009).

[4] Oriti (2014), Padmanabhan (2015).

[5] See Huggett and Wüthrich (2013a,b, 2018), Huggett (2017, 2018), and Wüthrich (2017, 2019).

The usual way of thinking of the architecture of the universe in the emergentist paradigm is that we start from the familiar world of everyday sense and peel back the layers revealing the physical structures that underlie the macroscopic surface of the world. Lower levels generate structures that scaffold higher levels, and as we descend to lower levels the structures become increasingly austere and increasingly unfamiliar. Colors and sounds get peeled away. The temporal asymmetries arise at an intermediate level, as a result of asymmetries in the distribution of matter. When the material contents of the universe are stripped away, what remains is a four-dimensional Lorentzian manifold in which space and time are united in a now familiar way. Below that level, we find lattices, or causal nets, or whatever one's favored theory offers for fundamental structures.

The presumption in this way of thinking is that we should be able to recover a concrete realization of general relativistic spacetime at a quite deep level, long before there is any thought of introducing agents. This program has encountered difficulties, both technical and conceptual, and there is some interest in reexamining the assumptions that are limiting how people are approaching the problem. Learning about how spatial perception works can make one less instinctively wedded to this way of thinking and also yield some insight into what space is (for us). In the first part of the chapter, I'm going to shed light on what it means to have the kind of immediate awareness that we have of space. In the second part, I'm going to draw some lessons for research in quantum gravity. The move toward ontologies in which space is not found at the fundamental level is one that shakes concepts at the very foundations of our thinking. The point of the chapter is not to provide definitive answers, or to push us away from any of the ongoing efforts to recover space in familiar ways. It is rather to introduce a little more sophistication about spatial perception into the discussion of quantum gravity in the interests of opening up other avenues of approach.

9.2 Part 1

9.2.1 Mach on the Construction of Phenomenological Space

When you open your eyes and look around, what you see is a three-dimensional space, external to you, in which objects distinct from your sensations are arranged.[6] What we are finding when we lift the veil and look at what is going on below the threshold of conscious awareness is that what is *given* to you in

[6] Sometimes, you lose your bearings, and your experience becomes a whirl of confusing patterns of sound and light. The difference between that disorienting whirl of sensation and the orderly world of objects in space is not in general a difference in the information coming in through the sensory pathways, but in what the brain does to organize that information.

perception is the prepared product of a lot of processing by the brain. The brain is working quite hard to integrate neural signals coming in through different sensory pathways into a unitary framework. In cognitive terms, space *is* that unitary framework. It provides a single frame of reference in which the objects of sight, sound, touch, and smell are jointly located, allowing us to share information across those pathways and use that information to guide movement.

Questions about the role of space in experience have a philosophical history running from Berkeley and Hume, through Kant, Strawson, and Evans.[7] Although it is not customary nowadays, there was a time when physicists read philosophers and thought that it was part of their job to give an account of the natural world that included the observer and accounted for aspects of human experience that are closely tied to interaction with the physical world. Einstein was engaged with philosophical thinking throughout his life. And Mach wrote a series of essays for *The Monist*[8] in which he discusses the nature, origin, and development of our concepts of space from points of view of sensory psychology, history, and physics. Although not right in all the details, Mach's essays were remarkably prescient.

There were three essays: I. On Physiological, As Distinguished from Geometrical, Space, II. On the Psychology and Natural Development of Geometry, and III. Space and Geometry from the Point of View of Physical Inquiry. In the first essay he discusses the difference between the 'physiological spaces' associated with the senses and what he calls 'geometric space'. By 'geometric space', he means what we would think of as physical space: an external dimension that contains the material bodies that are objects of visual, tactual, and auditory sensation. In the first essay, Mach argues that physical space is derived from 'physiological space', which is itself built out of atomic information delivered by the senses and motor activity.[9] The discussion proceeds as follows. Mach observes that animals have sense organs composed of one- or two-dimensional arrays of sensory elements (photoreceptors in the retina, for example, or tactile receptors in the skin). He says that such an arrangement is well suited to providing information of two kinds: (i) information carried by the response of a given element (a given tactile receptor, for example, might be geared to indicate pressure or temperature), and (ii) information provided by which element in the array is responding. The first tells us something about the condition of the thing we are getting information about; the second tells us where that thing is located. Mach's conclusion is that

[7] Berkeley (1988), Evans (1985), Guyer and Woods (1996), Strawson (1959).

[8] *The Monist* is an International Philosophical Quarterly. The papers came out between 1901 and 1903 and were published as *Space and Geometry* in 1906.

[9] There's some unclarity on a number of points in Mach's text. I've interpolated rather freely and apologize for any scholarly violence.

> [T]he perfect biological adaptation of large groups of connected elementary organs among one another is thus very distinctly expressed in the perception of space. (Mach 1906, 13)

According to Mach, an array of elementary sensory receptors makes available a 'physiological space' of a certain character—visual space, tactile space, auditory space, etc. About these he says:

> The physiological spaces of the different senses embrace in general physical domains which are only in part coincident. Almost the entire surface of the skin is accessible to the sense of touch, but only a part of it is visible. On the other hand, the sense of sight, as a telescopic sense, extends in general very much farther physically.... Yet, loosely connected as the different space sensations of the different senses may originally have been, they have still entered into connection through association... (Mach 1906, 15)

The physiological spaces associated with the senses are spaces in a purely formal sense. They consist of a set of elements with a relation defined on them. The fact that they form a space in the formal sense is no reason yet to think of them as having anything to tell us about physical space. To avoid confusion, I will refer to Mach's 'physiological spaces' as 'sensory manifolds', reserving 'Space' (capitalized) as a name for physical space. The pitches to which the human auditory apparatus is sensitive, for example, can be arranged along a one-dimensional continuum, from lower to higher. Receptors (hair cells) in the cochlea are arranged as a one-dimensional line of hairs sensitive to different pitches. Although vision presents itself as inherently Spatial, there is no more intrinsic reason to think of the sensory information delivered from the two-dimensional sheet of the retinae as Spatial than there is to interpret different pitches as indicating different locations in physical space. Nor is there any reason in general to think of the sensory manifolds associated with sight, touch, smell, and sound to have anything to do with one another.

What Mach suggests is that the sensory manifolds acquire a Spatial significance at the same time they acquire a significance for one another. They acquire that significance in virtue of being 'coordinated by association'. The associations in question are established through motor engagement. A creature that gets sensory information through different pathways and *acts* in the space that it gets sensory information about can establish robust, counterfactual-supporting relationships between the contents of different streams. It can use perceptual feedback on internally initiated movements to test relationships between what it sees and hears and feels.

The important upshot for our purposes is that Spatial import is not intrinsic; it is *acquired.* And it is acquired by coordination of the low-dimensional manifolds associated with the senses. Once the manifolds are coordinated with one another, a creature knows (immediately and non-inferentially and in a way that expresses itself in perception-guided movement) how to approach or retreat from things presented in its visual field. A dog knows how to get to the place from which the smell of meat is coming; a deer knows the direction in which to run to get away from the sound of danger. If you or I see a cup on a table in front of us, we know how to reach out and grab it, and that the steam and the temperature mean it will be hot against the lips. All of these implications are *extrinsic to the visual array*, but they are embodied in the Spatial significance which that array holds for us. They are products of the way in which that array is coordinated with other sensory and motor manifolds. And they are part of what it means to see something *as* located at a particular location in physical space.

The idea that sensory states acquire Spatial significance by being coordinated in a manner that brings with it a battery of sensorimotor skills is supported by much of what cognitive science has taught us about spatial perception. There is room for dispute about the interpretation of particular experiments, and arguments to be had about the philosophical significance, but the overwhelming evidence is pointing clearly to a connection between Spatial phenomenology and motor engagement.[10]

It is tempting to think of this as though the creature already has a notion of physical space (an internal map, or a representation of some kind)[11] and that the sensory manifolds acquire spatial significance by being associated with locations in *it*. But that is not the suggestion here. The suggestion is that the notion of physical space is the result of coordinating sensory manifolds, where coordination is a purely mechanical operation that precedes representation. Here's a little child's game we might use to illustrate. Suppose that you are given a bunch of patterns scribbled on papers of different sizes. You are not told anything about what they have to do with one another, but you find that there are regions of overlap that fit together in such a way that there is a way to piece them together by superimposing them onto one another, so that they form a coherent image of a three-dimensional object. And now you get the idea that what the patterns are is glimpses—from different angles—of the three-dimensional object rendered in the image. When you find parts of the pattern that seem to represent the same point, you put pins through them, in some cases overlaying multiple images

[10] For an excellent summary of the evidence, see Briscoe and Grush (2017). The article also contains a nice survey of philosophical disputes about its significance. The reference list there is up-to-date and extensive.

[11] I am using 'notion' in a way that is meant to span the space between the most rudimentary, pre-conceptual perceptual content and a fully articulated concept.

and utilizing the extra dimension you need to get all of the overlapping parts to fit together.

Once the construction is in place, the low-dimensional images appear as partial and redundant glimpses of the three-dimensional object, and they have a collective logic that is simpler than the sum of their parts. When I say they have a collective logic that is simpler than the sum of their parts, I mean that they have a collective geometry that is simpler than the conjunction of the patterns if those patterns are treated as independent. The power of this simplification expresses itself in the physics that would result if the object was put into motion. There is a drastic reduction of degrees of freedom in the assembled object that makes its physics much simpler than that obtained by treating each point on the diagrams as an independent degree of freedom. Once the object has been constructed—i.e., once you have 'solved for the object'—it would be natural and efficient for you to *see the patterns through the filter of this construction*. You would look at individual patterns for what they told you about the object, retaining information about the object, throwing away information about the patterns. You'd exploit the fixed connections between images embodied in the geometry of the object to bring information from one image to bear on another.

9.2.2 Coordination

Something like this process of construction, or triangulation, is going on in the mind as it weaves patterns across sensory manifolds into a coherent image of a three-dimensional world.[12] It is more complicated in a number of ways. It is not simply a matter of superimposing one image on another, but of weaving together streams of information about a world that is itself in flux. Identification of co-representing points is replaced with coordination of different sensory manifolds.[13] Although the contents are overlapping, the relationships among contents of streams are themselves changing (so, for example, if you have your hand on an unmoving object, and you turn your head, your visual array will change even though your tactual experience will stay the same; avert your eyes, turn your head, move your legs, and you disturb the connections between sensory arrays). The task

[12] There is a temptation to understand the kind of mapping in question in representational terms: we have a little map of egocentric space and the physiological spaces and we have little arrows representing mappings of the various sensory manifolds into the map of egocentric space. The problem with that is that—before any of this has any significance—the physiological spaces need to be coordinated with one another in a purely mechanical way. Once they are coordinated with one another mechanically, egocentric space appears a high-level virtual manifold stabilized out of coordination between the sensory manifolds. It is then made explicit in the form of mental images and concrete representations of space.

[13] Simplicity can be measured here in overall degrees of freedom.

for the mind is to piece all of this together into a coherent image of a single world, introducing as many degrees of freedom as it needs in order to accommodate variation over the streams individually, but utilizing regularities across streams to simplify the overall construction.

Just to give some sense of the difficulty of the task, consider a simplified example from Grush:

> Imagine a creature with two eyes, each of which can move in the creature's head. The creature has an arm with shoulder and elbow joints, which can bring the hand within the visual purview of the creature. There are a great many manifolds at work here. There is a 2-D manifold for each retina, a 2-D manifold for the orientation of each eye in the creature's head. There is a 2-D manifold for the orientation of the creature's head with respect to its body, and a 3-D manifold for the position of the creature's hand with respect to the torso (2 degrees of shoulder freedom, and one of elbow freedom). Let us also suppose that there's some bright object in front of the creature, which it can sense visually, as well as via touch. The project of c-coordinating the region available to vision and the region available to tactile sensation demands some sophisticated orchestration. Before they can be c-coordinated, some region of one must be identified with some region of the other. The only regions that will be stably identical are regions that are in a space centered on the creature's torso. All other regions will slide past and through each other as the creature moves this or that body part (for instance, as I move my head, my craniotopic space slides through my torso-centered space). So each must be stabilized to a reference frame anchored to the torso, as follows. The bright object will project an image onto each of the creature's retinae. However, the location of the object relative to the creature's head cannot be determined on the basis of the representations in the retinal manifolds, because the eyes may move while the head and object remain stationary, and this will change the position of the image on each retina. But, if given access both to the retinal images and to the orientation of each of the eyes in the head, the position of the object relative to the head can be fixed. That is, by appropriately s-coordinating the retinal and eye position manifolds, one can stabilize a 3-dimensional wedge anchored to the head. This 3-dimensional wedge is, as a set of ordered elements, a new manifold.[14] (Grush 2000, 68–69)

[14] See also Grush (2009, 311–345). Grush distinguishes 's-coordination' from 'c-coordination'. s-coordination involves simple identification of elements across manifolds. c-coordination involves recognition of additional degrees of dynamical freedom in the relationship between elements. In the text, I have used 'coordination' generically to cover both cases.

We're not yet finished. As Grush notes:

> This visual wedge is of little immediate use, however, because the head is free to move—something can remain at the same location in the wedge, while changing location relative to the torso and arms, if the head moves. This visual wedge needs to be stabilized with respect to the torso. This is done through stabilization with respect to a 2-dimensional manifold that provides information about the orientation of the creature's head relative to its torso, information provided by the muscles which control the position of the neck, mechanoreceptors which give information about the head's orientation, efferent copies of commands to move the head, and vestibular information. When this visual scene wedge is s-coordinated with the position of the head with respect to the body, then it becomes possible to stabilize a visual region with respect to the creature's torso. That is, for every pair (a, b) where a is an element in the visual wedge, and b is a specification of the orientation of the head with respect to the body, there is an associated point, an element in a manifold anchored on the torso. The region accessible to touch will be stabilized in an analogous manner, only though the s-coordination of rather different manifolds—limb joint angles, positions of stimulation on the skin surface, etc. When these two higher-order manifolds, visual and tactile, have been stabilized with respect to the creature's torso, it will then be possible to establish stable regions of overlap. These regions of overlap then underwrite the c-coordination of the higher order manifolds stabilized with respect to the creature's torso. The c-coordination of the 'visual space' and the 'tactile space' allows for the understanding that the felt thing and the seen thing seen are the same thing. (Grush 2000, 74)

The example illustrates how the physical act of coordinating sensory manifolds results in a higher-level virtual space, centered on the organism's body. This space is a three-dimensional manifold, anchored to the torso, which can serve as a common frame for all the lower-order (sensory as well as motor) manifolds that are coordinated with it. The result of the process is a collection of sensory manifolds coordinated and assembled on a platform that allows motion relative to one another, and that is itself moving through space. A physicist will think of this in terms of establishing relations between points in coordinate representations that are moving with respect to one another, and will think of the process of construction as a generalized best-matching procedure, where the only thing that is given is the manifolds. Topology (including dimensionality) of the embedding space is up for grabs.

And all of this suggests, again, that neither sensation nor action carry Spatial content intrinsically. The notion of Space, as a three-dimensional manifold that is the common arena of perception and action, is literally built by the coordination of sensorimotor manifolds. It arises first as a higher-level virtual space, represented only implicitly in the stabilization of the relations between lower-level manifolds, made explicit later in maps and mental images of ourselves as embodied agents in space. Not all sensory manifolds have Spatial significance because only some are coordinated with resulting higher-level manifolds. So, for example, retinal locations have Spatial significance, while pitches and temperatures (ordinarily) do not. Creatures must coordinate sensation and action through sensory systems and motor systems that are all connected to the torso on a variety of movable platforms (neck, shoulders, etc.) and a higher-order space centered on that common platform.[15] This platform is what we think of intuitively as our Point of View. The name is not entirely apt, because it is not really a point and it is not really a view. It is a frame of reference, with as many dimensions as the space it represents and all of the asymmetry needed to keep things aligned. And it is not tied to vision in any essential sense. What serves as the Point of View here is really the body as a whole.

Although the resulting construction sounds very complex if you try to think of it in terms of coordination among sensory surfaces, as soon as you realize that you can organize all of this complex information about the relationships between sensory and motor manifolds by just thinking in terms of information coming through sensory surfaces distributed over a body moving through Space, things snap neatly into place. Everything that we said about the complex network of evolving relationships among sensory streams is encoded in the view of oneself as an embodied agent seeing/hearing/touching/smelling a three-dimensional space through the sensors of a body whose movements one controls.

We can recover all of these relationships by thinking from the outside in, or we can consider the inverse problem that one would face working from the inside out. The point of starting from the inside and working our way out is to illustrate that constructing (a notion of) space is a hard problem. The inverse problem—retrieving the relationships among manifolds that it encodes—is an easy one. That is what makes that image a good user interface. Its role, in cognitive terms, is to explicitly represent the degrees of freedom that the part of the mind that manages sensorimotor connections needs to keep track of in order to get the body moving in the right way in response to the right stimuli. It does that in the simplest way it can, eliminating redundancy and suppressing detail not needed for the task. It is also, not incidentally, the representational space in which the sorts of higher-level management tasks that require slow, deliberate thinking (decision making and long-term planning) work. The part of the mind that performs those tasks is a very small part of the mind, from a cognitive science perspective, but the only one we know first-hand. It is the conscious part of the mind.

[15] Notice that the account of what it means for our experience to have Spatial content here does not appeal to topographic maps. All of this is to be understood functionally, without appealing to neural implementation.

Once the image of physical space has been constructed, it is not tied exclusively to any particular sensory modality. It doesn't matter whether you learned where an object is by sound or touch; you know where to look if you want to see it. Sensory information represented in a Spatial format is connected immediately and non-inferentially in a web of conditional expectations for visual, tactual, kinesthetic, and auditory experience. The knowledge built into that web of connections expresses itself as practical know-how; most notably (but not exclusively) the ability to use visual information to guide motor responses. A creature that sees the world through the lenses of this construction will know immediately and non-inferentially how to approach an object seen in the visual field, reach out and grab it, avoid an obstacle, and so on.

The connection between spatial phenomenology and sensorimotor ability is developed in action-oriented theories of perception. An early version is found in Gareth Evans:

> The subject hears the sound as coming from such and such a position, but how is this position to be specified? We envisage specifications like this: he hears the sound up, or down, to the right or to the left, in front or behind, or over there. It is clear that these terms are egocentric terms: they involve the specification of the position of the sound in relation to the observer's own body. But these egocentric terms derive their meaning from their (complicated) connections with the actions of the subject...
>
> Auditory input, or rather the complex property of auditory input which codes the direction of the sound, acquires a spatial content for an organism by being linked with behavioral output... (Evans 1985, 384–385)

The suggestion here is not to reduce perceptual Spatial content to a 'muscular vocabulary'. Nor is it to reduce the Spatial content of some perceptual modalities to that of one or more others (as there was for Berkeley, for example, who looked to reduce the spatial content of vision to that of touch and movement). It is rather that what it means for a creature to (for example) *see* space is that visual information is connected (for that creature) in a whole network of perceptual inputs and behavioral outputs. As Evans says:

> Egocentric spatial terms are the terms in which the content of our spatial experiences would be formulated, and those in which our immediate behavioral plans would be expressed. This duality is no coincidence: an egocentric space can exist only for an animal in which a complex network of connections exists between perceptual input and behavioral output. A perceptual input...cannot have a spatial significance for an organism except in so far as it has a place in such a complex network of input-output connections. (Evans 1982, 154)

9.2.3 Phenomenology

One's own experience (particularly visual and tactual experience) seems so intrinsically and inextricably Spatial that it is hard to separate what is *in* the sensory array from what is '*acquired by association*' (to use Mach's phrase). To make this phenomenologically plausible, it helps to look at examples of experience that has been artificially enhanced in a way that allows it to acquire Spatial content.[16]

Consider, for example, a device called a Sonic Guide, which is meant to provide blind people with a sense of distance so that they can navigate by sound. It comes with a gadget worn on the head that transmits a continuous high-frequency (inaudible) probe tone and picks up echoes of that tone with a stereophonic microphone. The objects in the user's vicinity reflect components of the probe sound in ways that depend on their size, distance, orientation, and surfaces.

The guide transforms echoes into audible sound profiles relayed to the user through earphones. The details don't matter for our purposes, but just to give a sense of how it works:

1. Echoes from a distance are translated into higher pitches than echoes from nearby (as a surface approaches a user, the sound it reflects will be translated into a tone that lowers as it gets closer).
2. Weak echoes are translated into lower volumes (as an object approaches, the wearer will hear a tone that increases in volume and gets lower in pitch. An object that grows but stays at a distance will get louder, without a change in pitch).
3. Echoes from soft surfaces (grass or fur) are translated into fuzzier tones, while reflections from smooth surfaces (glass or concrete) are translated into purer tones.
4. The left-right position of the reflecting surface is translated into different arrival times of the translated sound at each ear.

As Heil (1987, 238) says, the Sonic Guide[17]

> . . . taps a wealth of auditory information ordinarily unavailable to human beings, information that overlaps in interesting ways with that afforded by vision. Spatial relationships, motions, shapes, and sizes of objects at a distance from the observer

[16] Here, and throughout, we are assuming that sensory states have a phenomenology. The question is not what gives them a phenomenology, but what makes it *Spatial.*

[17] A similar device is the Tactile-Visual Sensory Substitution (TVSS). This is a device that doesn't use auditory information but comes with an array of tactile stimulators (small vibrators) worn by a blind subject on the stomach or back, and driven by a video camera usually worn on the subject's head. Each vibrator in the grid is driven or not, depending on the brightness of the video camera pixel to which it corresponds. Once they've become accustomed to navigating with the TVSS, subjects report ceasing to be aware of vibrations on one's skin and becoming aware rather of objects and events scanned by the camera. See Bach-y-Rita (1972).

are detectable, in the usual case, only visually. The sonic guide provides a systematic and reliable means of hearing such things.

People who have been using the device for a while and are competent with it report perceiving the objects in their environment directly. Although they are aware of pitches and volumes, there is no conscious inference or reasoning out what the environment must be like on the basis of pitches and volumes. The auditory experience of someone using the device from birth, who has never known space in any other way, has a Spatial content not different in kind from the spatial content that your visual states have for you. If asked what he hears, such a user says that he hears an object of such and such a size located over there, just as when you are asked to report what you see, you say you see a man at the door or a book on the table. Although the sound has a quality that he can focus on and describe, the significance that the sound has for him comes from its role in his sensorimotor economy and is described in terms of what it tells him about the locations of things in space—just as, although the visual array has a quality for us that we can focus on and describe, the significance it has for us comes from what it tells us about the location of things around us in space. We experience it as immediately and non-inferentially informative about the locations of things around us in space. Upon hearing a sound that represents (for him) an object located three feet from him and to the left, he would be prepared to immediately (and non-inferentially) orient toward it, point at it, or reach out to grab it, just as *seeing* an object three feet from *you* prepares you to do those things. If you or I put on the Sonic Guide and hear the same sounds, they would be just a pattern of sound signals, carrying no Spatial significance for us at all. There would be no question of using the thing to navigate, no chance of reaching out and grabbing an object at (for example) 24dB at middle C.[18] Gamers will find all of this very natural. They routinely

[18] Something similar can be said about his visual experience, if he suddenly gained sight. At first, his visual experience would have no Spatial significance. It would take some time for that whirl of light and color to become coordinated with the other sensory and motor manifolds. Once it did become coordinated so that he could reach out and grab an item in his visual field and use visual information to navigate, his visual experience would have as much Spatial content for him as ours does for us. Of course, in saying this, I'm taking an interpretational stance. One might accommodate all of the experimental data and deny (despite subjects' reports) that the Spatial phenomenology is immediate and non-inferential. Likewise, one might insist that with respect to visual phenomenology, that the Spatial content is at least to some degree innate. See Briscoe and Grush (2017) for discussion of the evidence. I stand by the interpretation, but the philosophical point doesn't hinge on it. Whether the Spatial content is hard-wired or inferred, the philosophical point is that it is causally mediated by physical processes between skin and skull, and a physical theory can reproduce the phenomenology without presuming it mirrors external structure. There has to be some correspondence, but the nature of the correspondence is dictated by the role of visual space in coordinating sensorimotor behavior. Does having spatial phenomenology *consist* in possessing this pattern of sensorimotor skills? Is it merely *produced* by that pattern? I leave that question unsettled, because we don't need to settle it for our purposes. Note also that one might deny that our Spatial experience is as unified as I've supposed. For example, one might deny that auditory and visual space has the kind of seamless integration I've supposed. Again, here, these controversies leave the philosophical point intact. The reason for

report ceasing to be aware of their own bodies as they acquire sensorimotor skills in alien environments. They aren't aware of the movements of fingers on a keyboard or hands on a joystick. They are aware, rather, of jumping or flying or propelling a virtual body through a virtual space. In some cases that virtual space is three-dimensional, but it can also be four-, five-, or six-dimensional. And it is not physically realized (as an external structure in which the gamer and game are somehow contained): it is merely represented in the two-dimensional expanse of the screen. The way to describe this is that the gaming context creates an immersed environment in which visual feedback on the muscular sensations that we ordinarily associate with finger and hand movements becomes associated with movements of the virtual body on the screen. In that environment, it is experienced as movements of the body they control, in the space in which it lives. Sensorimotor connections are locally and artificially hijacked into the virtual world.

The suggestion of all of this is that the Spatial content of our own experience is acquired by being caught up in a coordinated web of sensorimotor manifolds. It is the coordination of these manifolds that allows them to inform one another and also guide action. The difference between you and I, on one hand, and the person who has learned to use the Sonic Guide, on the other, is that his auditory sensations have gotten caught up in a web of sensorimotor connections in virtue of being coordinated with the other manifolds. This is what 'learning to use the device' amount to. At the same time that his auditory experiences have acquired Spatial content, his Spatial concepts have acquired a sound profile. An auditory experience (and more generally, the elements in a sensory manifold) acquires Spatial significance by being associated with locations in the higher-level space stabilized out of the coordinations among manifolds. And the elements in the higher-level manifold get their content from the entire web of sensory and motor manifolds that go into its construction.

That higher-level manifold is made explicit internally in the form of a mental image that can serve as input to conscious processing. It is made explicit externally when we make maps and the like.[19] We learn to picture space in our heads in part by looking at pictures and maps. By the time we are in a position to raise the question of which comes first, we are well past the stage at which they can be separated. As we are educated our Spatial concepts become more abstract,

presuming that the familiar three-dimensional space of everyday sense had to be recovered as an external structure is that we seemed to have immediate awareness of it as such. If space is not even phenomenologically unified, that weakens the *prima facie* case against the presumption.

[19] Older space-representing technologies—like maps—tend to represent space in a frame-independent manner, a manner that is invariant under transformations of point of view. Many new dynamical digital technologies (like GPS) represent space in terms that are relativized to (and vary with) the user's location. What gives these representational technologies Spatial content is (plausibly) the way in which they are caught up in the sensorimotor behaviors of their users.

systematic, and articulated. How much of that additional learning filters into the way that space is presented in perception (in cognitive science terminology, how penetrable perceptual contents are to theoretical knowledge) is disputed. What is agreed on all sides is that perceptual contents directly represent things as having locations in a Spatial frame of reference anchored to our bodies. That is the sense—and I am suggesting, the only sense—in which we see space. It is the sense—and the only sense—in which we have immediate awareness of the visible and tangible *reality* of space.

9.2.4 The Upshot

There are a great many unanswered questions about spatial perception, i.e. the process that transforms sensory signals into Spatially structured representations of the world. The point of the above is to give some sense of the rich body of scientific research that is bringing that process into view. The fact that we have (phenomenologically) immediate awareness of the visible and tangible reality of space makes us instinctive realists about space. From an untutored point of view, space is the dimension that houses objects that exist outside of our own minds; the things that we see, touch, hear, and smell. We are as certain of its existence as we are of the external world.

From the side-on perspective adopted above, however, the presumption that our perceptual states give us a direct, unfiltered view of reality as it is appears rather naïve. Perception is an embodied process. The products of perception were selected for their role in the organism. That role is not to provide a mirror of reality, but to provide a user interface serving the part of the mind that manages sensorimotor connections. Whether, and in what sense, reality itself is spatially structured is the question of what the world has to be like to produce that pattern of sensorimotor connections. That is the question that physics has to address. The burden that the Spatial appearances place on a theory is to reproduce that pattern of connections.

In the remainder of this chapter, I will look at how we might bring these lessons to bear on research on quantum gravity.

9.3 Part 2

By and large, research in quantum gravity has been shaped by a presumption that the task of recovering space demands finding space(-time) as a concrete emergent structure in the world, prior to the introduction of agents. There are various reasons for this presumption. One is that physics tends toward conservatism in its methods. Physicists are familiar with methods that involve recovering a theory in some limit, with approximations. It is not a bad policy to start with what you know and explore new methods when that doesn't work. A less benign reason is that there is a strong disciplinary resistance to introducing any talk of observation or experience into physics.

9.3.1 Maudlin's Warning: Stay Away from the Observer

Often passive and quiet, that resistance is explicit in an important paper by Tim Maudlin that has gotten a good deal of (deserved) attention from those working in quantum foundations and quantum gravity.[20] Maudlin suggests that 'naïve realism' about space might be a *sine qua non* of physics. He argues is that a theory that does not contain local beables—concrete material structures like measuring instruments occupying a public arena whose properties could serve as touchstones for our joint experiences, and whose observable properties could provide the evidence for our theories—faces a big challenge connecting evidence to theory. The argument goes like this:

> Because of the mediating place of local beables, classical physics could be tested without mention of the mind-body problem or the problem of connecting claims about experience with physical descriptions: the evidence, after all, was stated not in the language of experience but in the language of local physical facts (e.g. that the rocks hit the ground together). If the local beables and the locations are removed from the physical ontology, it is hard to see how evidential contact with the world is to be made *except* at the level of conscious experience.
>
> (Maudlin 2007, 3160)

Maudlin's worries are directed at a theory that contains no spacetime, even at the level of emergent structure.[21] The reasoning is that if there is no spacetime, there are no local beables. That means we are forced to describe the evidence in phenomenological terms, and that leads into a philosophical quagmire. Classical physics didn't face such difficulties because it contained local beables, and it was precisely by avoiding those problems that it made progress. He writes:

> Some connection is supposed to be recognizable between [a given] physical description and our conscious *experiences*, even though we have nothing like a precise vocabulary with which those experiences, as such, can be described. [The]

[20] Maudlin (2007, 3151).

[21] See also Lam and Esfeld (2013), and for a response de Haro and de Regt (2018).

> mediating role of local beables obviated all these problems: there is no doubt a question about how, when we look at a rock with a certain shape, a conscious experience of certain sort arises, but classical physics could put that question off for another day (which perhaps would never come). All classical physics needs is the belief that experiences *as of* a rock of a certain shape typically *are* experiences of a rock with that shape, and the physics could take care of the rock. It is hard to see even how to begin if the physics has, in its own terms, to take care of the *experiences*. (Maudlin 2007, 3160–3161)

Later in the paper he goes on the offensive and puts the challenge in terms of a threat that Huggett and Wüthrich have characterized aptly as involving a form of empirical incoherence.[22] The worry is that if our evidence takes the form of observations of the values of local beables, and the theory entails that there are no local beables, then it undermines the very evidence that was used to establish it. The theory becomes unstable in the sense that one can't simultaneously assert the theory and hold that it is supported by the evidence. One can distinguish here a strong position and a weak position. The strong position is that the existence of local beables is a non-negotiable constraint on ontology, on pain of empirical incoherence. The weak position is that if there are no local beables, we have lost a connection between the evidence and the theory, and the only way to recover the link is to put the observer back into the picture. Maudlin actually defends the weaker position, but he wraps it in such passioned excoriation against the introduction of talk of experience or observation into physics—with gestures at the failure of positivism and the mess of the mind-body problem—that it is clear that he thinks one would have to be a fool to defend a theory with no local beables.

It is hard to tell how seriously Maudlin intended the reference to the mind-body problem, but it is worth addressing because the whole paper is pervaded by a sense—widely shared by physicists—that whenever we start talking about experience, we are in the province of a philosophical morass. One of the nicest things that has happened in the philosophical discussion of consciousness in the last twenty years is the separation of philosophical and scientific questions about the mind. A lot of the credit goes to David Chalmers who separated the mind-body problem into Hard and Easy Problems. Easy Problems, in Chalmers' terminology, are explanations of things like the capacity to discriminate, categorize, and react to stimuli from the environment; the ability of a system to access its own internal states and to report them if queried; and the deliberate control of behavior. These are all abilities that can be functionalized and studied scientifically. The Hard Problem, in Chalmers' terms, is to give an account of the nature of phenomenal

[22] Huggett and Wüthrich (2013b).

consciousness. What makes the Hard Problem hard, according to Chalmers, is that it demands an explicit constitutive account in physical terms of what it takes for some state or mental event to be conscious, and there is a collection of arguments that seem to show that whatever one offers in the way of a functional characterization of our own conscious states will fail as a constitutive account. Being conscious is not (according to Chalmers) a functionally characterizable property.

Chalmers' primary interest was in protecting the philosophical problem about the nature of consciousness from scientific encroachment. In doing so, however, he freed the sciences of the mind to study perception, cognition, and other mental phenomena without getting hung up on worrying about what consciousness is. The scientist can make conscious mental events an object of study, identifying them by a functional definition (e.g. introspective accessibility, availability for global broadcast)[23] and examining their internal role in human cognitive architecture and their causal connections to events in the environment. So long as she is not offering a constitutive account of what consciousness is, she is perfectly within her rights to ignore the hard problem by saying that insofar as consciousness matters to the cognitive/behavioral economy of an organism, it can be functionalized and studied scientifically. Insofar as it can't be functionalized and defined in physical terms, it doesn't make a difference to the cognitive/behavioral economy and can be safely disregarded.[24]

What it means for physics is the possibility of reincorporating the observer explicitly as a part of the physical system that is implicated in the production of the experience.

9.3.2 Putting the Observer Back In

Observer avoidance might have been a good methodological strategy when the status of space was secure and the scientific study of perception wasn't very developed. But in a setting in which physics is exploring ontologies in which space doesn't appear at the fundamental level, and we have a large and growing body of scientific research on how spatial perception works, it makes no sense. Like it or

[23] Introspective accessibility is the most natural way of identifying conscious states from a pre-theoretical point of view, but global broadcast, availability to deliberative processes, and other criteria have been proposed in the context of particular functional accounts. To say that the scientist identifies conscious events by an extensional definition (e.g. here, in terms of functional role) is to say that she has some criterion that picks them out by a definite description, in the way that we might pick out sugar as 'the dominant ingredient in lollipops' without offering that as a constitutive account of what sugar is.

[24] This is not to endorse Chalmers' dualism, but simply to allow us to study the cognitive/behavioral economy of the system without trying to answer the problem of consciousness in Chalmers' own terms.

not, the ultimate source of information about the world for each of us is perceptual experience. It is unavoidably a part of the physical chain that connects ontology to evidence and pushing it out of view doesn't make it go away.

Nor is there a need to push it out of view. What happens between skin and skull are physical processes. It became clear quite early on in the scientific study of the mind that it was most natural to speak of those processes not in neural terms, but in an information-theoretic vocabulary. Cognitive science—which models the mind as a system that processes and stores information, bringing it to bear on the determination of behavior—was born and developed much more quickly than it would have if we had to wait on an understanding of the underlying neural properties.[25] There has been an enormous amount of progress in the last half-century in studying perception in this way. In cognitive scientific terms, perception is modeled as a physical process that transforms signals coming in through sensory pathways into introspectively accessible states that represent the world as being a certain way. The process that produces those states is enormously complex and happens mostly below the threshold of introspective awareness. It can be studied using a battery of methods from behavioral experiments, to brain imaging and computational modeling. Bringing it into view makes it clear that the choices that Maudlin seems to present us with—either take perceptual experience at face value or fall into empirical incoherence—are not the only ones available. The considerations from Part 1 were intended to turn the tables on Maudlin's style of realism, by suggesting (at least implicitly) that taking experience *at face value*—if that means assuming that our perceptual states mirror the world—may be naïve. A cognitive scientific look at perception, which pays attention to how perceptual states are produced and the role they play in the organism reveals that experience is neither revelation nor illusion. It does not produce representations that simply mirror the environment, but it does carry information about the world.

The way to exploit that research in the service of theories of quantum gravity may be to switch from asking 'How much structure does a theory have to have in order to recover *physical space* (i.e. to recover spacetime as a concrete structure in the external environment)?' to asking 'How much structure does a theory have to have to recover *phenomenological space* (i.e. to recover the Spatial content of human experience, or if you like, the Spatial *appearances*)?'. We model the observer as an information-processing system coupled to the world through sensory pathways, capable of initiating action through motor pathways, and getting sensory feedback on its actions. And we ask: what would the world have to be like to produce the whole structure of relationships between sensory and motor manifolds

[25] That is fortunate for our purposes, because it allows us describe observation without using spatiotemporal vocabulary. We need only attribute to the world a minimal causal structure: enough to support both the sensory signals coming in and the information-processing itself.

that get organized by the mind into the image of ourselves as agents living in a three-dimensional world? So long as a theory reproduces the relations between these manifolds, phenomenological space will take care of itself.[26]

Every proposal for a theory of quantum gravity will have to reproduce spatial experience in its own terms. One way is to recover space as an external structure. But another is to do it without reifying space. The latter suggestion might be a natural pairing with the causal set approach, for example, and it implements the relational strategy associated most prominently with Carlo Rovelli. The relational strategy, as Oriti says so nicely here:

> takes on board the main lesson of GR [general relativity], and it rephrases it in a way that immediately suggests a tentative solution: there is no time and no space, but only physical (imperfect) clocks and rods. The strategy amounts to identifying internal degrees of freedom of the complete system composed of metric and matter fields that can be used as approximate rods and clocks to parameterize the spatial relations and temporal evolution of the remaining degrees of freedom.[27] (Oriti, this volume)

All that we need to do to adapt this to the present setting is replace 'clocks and rods' with 'information-processing agents'.

Ultimately what physics has to recover are the spatial appearances and that this can be done without recovering space as an external structure (and also without lapsing into idealism) is an important observation. Is it likely that this will help research in quantum gravity? Since the 'loss of space' is a quantum phenomenon, it concerns only structure at the very small scale and is washed out rather quickly by decoherence. So there may be in the end good reasons to expect space to emerge directly before the introduction of agents. It is important nevertheless to have the possibilities on the table, to be clear on what empirical adequacy demands, and most importantly to establish that where exactly the line between internal and external structure lies is an *empirical* question, a question of *physics*. If space is recovered as an external structure, it is because the dynamical laws deliver it in a natural way, not because we can't do physics without space.

[26] The pattern of relationships between sensory and motor manifolds is structure in the signal, and it carries information about the world. The job of a theory of quantum gravity is to say how the world produces that signal. The manifest image of the world—the image that the mind prepares and presents in experience—is the message that a mind extracts from the signal. The recommendation to quantum gravity theorists is to ignore the message: focus on recovering the signal.

[27] Note also that the relational strategy has an answer to Maudlin's challenge that it is not enough that a theory allows for the emergence of structure isomorphic to spacetime. There must be something internal to the theory that makes that structure salient. In this setting, the structure has salience for an observer coupled to it in the right way. It has relational salience, not absolute salience. See also Rovelli (1991, 297–316); Gambini and Porto (2001).

So ultimately, it may be a primarily philosophical point that is being made here. We want to separate questions *for physics* from 'presuppositions of physical inquiry', or 'conditions of the possibility of physics', or anything like that. From a philosophical point of view, it is worthwhile to break down the idea that realism about physics demands a particularly naïve form of realism about Space. Seeing in detail what is going on in the mind to produce Spatial appearances tells us something about what Space ultimately is for us, and about what the visible and tangible reality of Space really amounts to. There is an intuitive and psychologically basic association between Space and externality. It is an association elevated to an analytic connection in much contemporary metaphysics where being 'physically real' is often identified as being 'located in space'. The material in Part 1 gives us some insight into why the association is there, while also showing us why physics need not respect it.

9.4 Conclusion

The idea that the mind weaves a conception of the external world out of regularities in experience is an old one in philosophy. In the Western tradition Berkeley's *A New Theory of Vision* is the *locus classicus*, and it was clearly in Mach's mind when he was writing the essays in *Space and Geometry*. Although that idea is associated with idealism, there is nothing essentially idealistic about it. Any regime that treats perception as a physical process has to recognize that, when information about the world crosses the threshold of the body, it has been transformed by the separation into different sensory streams and the mind faces a rather complex task in trying to produce an explicit model of (the causal structure of) its source.

This has been an introduction to research that shows Show it actually does that that holds two upshots for research in quantum gravity. The immediate lesson is that it raises the possibility of a program that tries to recover spatiotemporal experience without reifying space, by letting the mind supply some of the structure that organizes sensory information into a spatial format. The more amorphous, but more important, lesson is that perhaps it is time to bring the observer back into physics. Abner Shimony saw it as one of the central tasks of a completed science to bring metaphysics and epistemology together into a mutually supporting package; the fundamental ontology of our theory should allow us to recover our experience, and our experience in its turn should be a genuine source of information about the world. He emphasized, however, that the information that experience carries about the world is itself a matter for scientific illumination.[28] And he saw that when

[28] Shimony (1993). There is nothing certain about it, of course. It is all science, which means that it is experiment and evidence-driven conjecture. Right out of the gate we give away the prospect of the foundationalist dream of building our view of the world by logically sound inferences on epistemically secure foundations.

physics reaches maturity, the observer will have to appear explicitly as part of its subject matter. Only then will the circle be closed.[29]

References

Bach-y-Rita, P. (1972). *Brain Mechanisms in Sensory Substitution*. New York: Academic Press.

Berkeley, G. (1988). *Principles of Human Knowledge and Three Dialogues*. London: Penguin UK.

Briscoe, R. and R. Grush (2017). Action-based theories of perception. In E. N. Zalta (Ed.), *The Stanford Encyclopedia of Philosophy* (Spring 2017 ed.). Metaphysics Research Lab, Stanford University. https://plato.stanford.edu/archives/spr2017/entries/action-perception/.

de Haro, S. and H. W. de Regt (2018). A precipice below which lies absurdity? Theories without a spacetime and scientific understanding. *Synthese*, 1–29. https://doi.org/10.1007/s11229-018-1874-9.

Einstein, A. and Born, M. (2004). *The Born-Einstein Letters*. London: Palgrave Macmillan.

Evans, G. (1982). *The Varieties of Reference*. Oxford: Oxford University Press.

Evans, G. (1985). Things without the mind. In *Collected Papers*, pp. 249–290. Oxford: Oxford University Press.

Gambini, R. and R. A. Porto (2001). Relational time in generally covariant quantum systems: Four models. *Physical Review D 63*(10), 105014.

Grush, R. (2000). Self, world and space: The meaning and mechanisms of ego- and allocentric spatial representation. *Brain and Mind 1*(1), 59–92.

Grush, R. (2009). Space, time and objects. In J. Bickle (Ed.), *The Oxford Handbook of Philosophy and Neuroscience*, pp. 311–345. New York: Oxford University Press.

Guyer, P. and A. Woods (1996). *The Cambridge Edition of the Works of Immanuel Kant*. Cambridge: Cambridge University Press.

Heil, J. (1987). The Molyneux question. *Journal for the Theory of Social Behaviour 17*(3), 227–241.

Huggett, N. (2017). Target space ≠ space. *Studies in History and Philosophy of Modern Physics 59*, 81–88.

Huggett, N. (2018). Spacetime 'emergence'. In E. Knox and A. Wilson (Eds.), *Routledge Companion to Philosophy of Physics*. London: Routledge. https://www.routledge.com/The-Routledge-Companion-to-Philosophy-of-Physics/Knox-Wilson/p/book/9781138653078.

[29] I'd like to thank Carlo Rovelli and David Albert for comments on an early draft. I owe them both enormously. Many thanks also to Nick Huggett and Christian Wüthrich and all the participants in the Chicago area summer school on the emergence of spacetime in quantum gravity. I owe a special debt to Rick Grush, who taught me to think about space in a completely new way, and whose work I rely on heavily here.

Huggett, N. and C. Wüthrich (2013a). The emergence of spacetime in quantum theories of gravity. *Studies in History and Philosophy of Modern Physics 44*(3), 273–275.

Huggett, N. and C. Wüthrich (2013b). Emergent spacetime and empirical (in)coherence. *Studies in History and Philosophy of Modern Physics 44*(3), 276–285.

Huggett, N. and C. Wüthrich (2018). The (a)temporal emergence of spacetime. *Philosophy of Science 85*(5), 1190–1203.

Kiefer, C. and C. Weber (2005). On the interaction of mesoscopic quantum systems with gravity. *Annalen der Physik 14*(4), 253–278.

Lam, V. and M. Esfeld (2013). A dilemma for the emergence of spacetime in canonical quantum gravity. *Studies in History and Philosophy of Modern Physics 44*(3), 286–293.

Mach, E. (1906). *Space and Geometry in the Light of Physiological, Psychological and Physical Inquiry*. Chicago, IL: Open Court Publishing Company.

Maudlin, T. W. (2007). Completeness, supervenience and ontology. *Journal of Physics A: Mathematical and Theoretical 40*(12), 3151.

Oriti, D. (Ed.) (2009). *Approaches to Quantum Gravity: Toward a New Understanding of Space, Time and Matter*. Cambridge: Cambridge University Press.

Oriti, D. (2014). Disappearance and emergence of space and time in quantum gravity. *Studies in History and Philosophy of Modern Physics 46*, 186–199.

Oriti, D. (2021). Levels of spacetime emergence in quantum gravity. In C. Wüthrich, B. Le Bihan, and N. Huggett (Eds.), *Philosophy Beyond Spacetime*. Oxford: Oxford University Press. This volume.

Padmanabhan, T. (2015). Emergent gravity paradigm: Recent progress. *Modern Physics Letters A 30*(03n04), 1540007.

Rovelli, C. (1991). What is observable in classical and quantum gravity? *Classical and Quantum Gravity 8*(2), 297–316.

Shimony, A. (1993). *The Search for a Naturalistic World View*, Volume 1. Cambridge: Cambridge University Press.

Strawson, P. F. (1959). *Individuals. An Essay in Descriptive Metaphysics*. London: Routledge.

Teitelboim, C. (1982). Quantum mechanics of the gravitational field. *Physical Review D 25*(12), 3159.

Wüthrich, C. (2017). Quantum gravity from general relativity. In E. Knox and A. Wilson (Eds.), *Routledge Companion to Philosophy of Physics*. London: Routledge.

Wüthrich, C. (2019). The emergence of space and time. In S. Gibb, R. F. Hendry, and T. Lancaster (Eds.), *The Routledge Handbook of Emergence*, pp. 315–326. London: Routledge.

10

The Measurement Problem for Emergent Spacetime in Loop Quantum Gravity

Richard Healey

Abstract

Any quantum theory of gravity faces the measurement problem. Carlo Rovelli sees his relational interpretation as offering a solution to this problem when applied to his favored loop quantum gravity (LQG). I examine the prospects of Rovelli's relationalism in LQG. In LQG it is not clear what physical systems there are at a fundamental level with no spacetime. But implementing Rovelli's relational interpretation in the context of LQG requires an account of interaction and a model of observer systems whose relational states represent determinate outcomes. Even if no such account is forthcoming at a fundamental level in LQG, it might still be available in a limit at which spacetime has emerged. But to use this account effectively to address the measurement problem it would be necessary to reconcile the observer-relativity of measurement outcomes with a basic norm of scientific objectivity.

10.1 Introduction

It is widely believed that a quantum theory of gravity will not take spacetime to be fundamental, but that our familiar spacetime structure will rather emerge from something more fundamental in a suitable limit of the theory. Any putative quantum theory of gravity must address the quantum measurement problem, and one promising strategy for addressing that problem involves a relational view of states in quantum theory.

Several different ways of implementing this strategy have now been proposed. These include Rovelli's relational interpretation (1996), QBism (Fuchs 2010, Fuchs, Mermin, and Schack 2014), and a pragmatist approach (Healey 2012, 2017). No one has done more to solve the conceptual as well as technical problems posed by a projected quantum theory of gravity than Carlo Rovelli. Anyone attempting to implement a relational strategy toward the measurement problem in

Richard Healey, *The Measurement Problem for Emergent Spacetime in Loop Quantum Gravity* In: *Philosophy Beyond Spacetime: Implications from Quantum Gravity.* Edited by: Christian Wüthrich, Baptiste Le Bihan, and Nick Huggett, Oxford University Press. © Richard Healey 2021. DOI: 10.1093/oso/9780198844143.003.0010

quantum gravity has much to learn from Rovelli's path-breaking work in applying his relational interpretation to covariant loop quantum gravity (LQG). So what follows should be viewed as an attempt to learn what I can from a critical but deeply sympathetic assessment of that work.

The chapter proceeds as follows. Section 10.2 examines how Rovelli's relational quantum mechanics applies to theories with a fixed background spacetime. This restriction is lifted in section 10.3, introducing LQG—Rovelli's own favored approach to a quantum treatment of the gravitational field. The measurement problem is first discussed in section 10.4, while section 10.5 further investigates the application of Rovelli's relational view of quantum theory to LQG. In section 10.6 all these pieces are put together to critically assess Rovelli's use of his relational interpretation to address the measurement problem for LQG. The final section 10.7 draws the morals of this story for alternative relational approaches to quantum theory when applied to the emergence of spacetime from a putative quantum theory of gravity.

10.2 Rovelli's Relational Quantum Mechanics

In his 1996 paper, Rovelli expressed the key idea behind his relational quantum mechanics as follows (italics in the original):

> *Quantum mechanics is a theory about the physical description of physical systems relative to other systems, and this is a complete description of the world.*
> (Rovelli 1996, 1650)

He makes his view clear that what the physical description is relative to is itself simply some other physical system:

> Observers are not 'physically special systems' in any sense. (1654)

Just as 'the observer' to which velocities must be relativized in Galilean relativity may be any physical object (such as a table lamp), so also 'the observer' to which the state of a physical system must be relativized in quantum mechanics may be any physical system (such as an electron).

To understand what Rovelli's key idea comes to, we need to ask how quantum systems are to be described physically, and what are the physical systems.

In physics, a system is usually described by assigning it a state: in quantum theory one expects this to be a quantum state, represented mathematically. But just how, and how completely, this mathematical representation describes the system's physical properties is not transparent—especially when the description is taken to be relative to other systems. I address Rovelli's account of the descriptive role of quantum states in the following subsection.

10.2.1 What Is a State?

In ordinary quantum mechanics, the state of a system is represented by a vector in or density operator on Hilbert space. It is common to attribute a determinate value q_i to quantity q on a system if application of the Born rule to that state yields probability 1 that observation of q would reveal value q_i: the eigenstate-eigenvalue link strengthens that conditional to a biconditional by adding that this is the *only* circumstance in which a quantity has a determinate value.

Rovelli (1996, 1997) endorses the eigenstate-eigenvalue link, though only for states relativized to a fixed 'observer' system. But he also proposes to replace this notion of the state of a system with the notion of the information that a system has about another system.

Given three physical systems *S*, *O*, *P*, by "... '*q* has a value [on *S*] relative to *O*', we mean 'relative to *P*, there is a certain correlation in the *S* and *O* states', or, equivalently, '*O* has information about *q*'" (Rovelli 1997, 10). But "... there is a key irreducible distinction between *P*'s knowledge that *O* has information about *q* and *O*'s knowledge of *q*" (Rovelli 1997, 10). In this situation, does *O* *have* knowledge of (the value of) *q*?

If the term 'knowledge' has its normal sense, then this question has an easy answer ('No') in a case where *O* is a simple quantum system such as an electron that has no capacity for knowledge. But one would expect it typically to have the opposite answer ('Yes') when *O* is an alert, competent, human experimentalist. Indeed, *denying* that a human observer can know the value of *q* leads to a form of skepticism that removes much if not all evidential support for quantum mechanics.

However, Rovelli cannot consistently allow that such knowledge contributes to a complete description of the state of the world.

> ... the incorrect notion at the source of our unease with quantum theory ... is the notion of true, universal, observer-independent description of the state of the world. (1996, 1637)

So the value of a quantity on *S* must be relativized to *O*, and if *O* is a human observer then his or her knowledge of its value cannot contribute to our knowledge of the observer-independent state of the world!

> ... when we say that a physical quantity takes the value *v*, we should always (explicitly or implicitly) qualify this statement as: the physical quantity takes the value *v* with respect to the so and so observer. (1648)

10.2.2 Systems, Interactions, and Processes: Quantum Mechanics

For Rovelli, even relational descriptions are not always available:

> A quantum description of the state of a system *S* exists only if some system *O* (considered as an observer) is actually 'describing' *S*, or, more precisely, has interacted with *S*. (1996, 1648)

Systems in quantum mechanics are particles or compounds of them: the distinction between simple and compound is made contextually. The state of a system is represented in its Hilbert space, on which is defined a Hamiltonian determining its undisturbed (Schrödinger picture) evolution.

The occurrence of an interaction between two systems is represented by a non-zero interaction Hamiltonian on their tensor product Hilbert space: this may be arranged to have a specific form in a von Neumann 'measurement' that effects a correlation between the initial state of a system *S* and the final state of another system *O*: I exhibit this form in section 10.5.

So in quantum mechanics the history of a system consists of processes of undisturbed evolution punctuated by episodes of interaction. It is represented mathematically by a quantum state that evolves unitarily (in the Schrödinger picture), though only between interactions with other systems is this state assigned to that system alone. For Rovelli, all states are defined relationally, because a state assignment to *any* system is relative to an observer system *O* that has interacted with it.

10.2.3 Systems, Interactions, and Processes: Quantum Field Theory

In Lagrangian quantum field theory on a fixed background spacetime, systems are quantum fields such as the photon, electron, and Higgs fields. Given a decomposition into positive and negative frequencies, the state of a field may be represented in a Fock space: its free evolution is represented by a continuous family of unitary operators on this space. Particles and classical fields may then also be counted as (emergent) systems corresponding to excited states in the associated Fock space.

The occurrence of an interaction between two quantum fields is represented in the total Lagrangian (density) by a non-zero interaction term coupling them.

A function of a *single* field operator in the total Lagrangian can imply non-linearity of its resulting Euler-Lagrange equation, indicating a *self-interaction* of that field (as in the gluon field in quantum chromodynamics). To either kind of interaction term corresponds an interaction Hamiltonian whose non-zero value signals the occurrence of an interaction, just as in quantum mechanics.

The paucity of fundamental interactions[1] leaves no freedom to engineer episodic von Neumann 'measurements' at will by designing novel fundamental fields. If measurement is to be modeled as a unitary process within quantum theory, this cannot be at the level of fundamental fields. But one may perhaps model von Neumann measurement interactions between *effective* quantum fields that emerge from more fundamental fields in an appropriate limit.

10.3 Loop Quantum Gravity

In LQG there is no background spacetime. So, unlike the photon or electron field, the quantum gravitational field cannot be modeled by an assignment of operators to points or regions of spacetime. Moreover, in pure LQG there are no other fermion or boson fields with which the gravitation field might interact.

Since the non-linearity of the Einstein field equations of general relativity at least intimates non-linearity of field equations for a quantized gravitational field, one might think of this field as self-interacting. But what could it be for two systems to interact in LQG?

Rovelli and Vidotto (2015, 52) say this about Rovelli's (1996) relational quantum mechanics (where LQG is not mentioned):

> . . . a process is what happens between interactions.

But they say this relational interpretation may be used in the context of quantum gravity, where

> . . . a process is not *in* a spacetime region: a process *is* a spacetime region.
>
> (2015, 52)

10.3.1 Systems, States, and Interactions in Loop Quantum Gravity

We saw how Rovelli handled the notion of a state in his relational quantum mechanics. How does this work in LQG? According to Rovelli and Vidotto (2015, section 2.4.3), the relational interpretation of Rovelli (1996) can be used in the

[1] The Standard Model Lagrangian includes just fundamental strong and electroweak interactions. That leaves gravity as the only other fundamental interaction.

context of quantum gravity. But that section leaves it unclear how to apply this relational interpretation to LQG. We are told that

> States are descriptions of ways a system can affect another system. (2015, 52)
>
> A state is not somewhere in space: it is the description of the way two processes interact, or two spacetime regions passing information to one another. (2015, 52)

It seems that the systems that interact in LQG are spacetime regions, themselves thought of as processes.

> A spacetime region *is* a process: a state is what happens at its boundary. (53)

However, the boundaries marking the division between processes are not fixed, but arbitrarily drawn in spacetime:

> The physical theory is therefore a description of how arbitrary partitions of nature affect one another. (53)

This is possible because

> As noticed, a remarkable aspect of quantum theory is that the boundary between processes can be moved at will. Final total amplitudes are not affected by displacing the boundary between 'observed system' and 'observing system'. The same is true for spacetime: boundaries are arbitrarily drawn in spacetime. (53)

Total amplitudes are related to processes because the process that is a spacetime region

> is actually a Feynman sum of everything that can happen between its boundaries. (52)

The total amplitude $\langle W|\Psi\rangle$ associated with the (quantum) state Ψ of the gravitational field on the boundary $\Sigma = \partial M$ of a spacetime region M may be expressed as a Feynman path integral of the field in M (2015, 50f). Assuming this is what the authors are referring to by the phrase 'Feynman sum', we should be able to conclude that the process that is a spacetime region M is simply this total amplitude $W = \langle W|\Psi\rangle$.

But now we have a problem, since an amplitude is a complex number, while a spacetime region is presumably not any kind of number but something in the physical world (perhaps a physical object, perhaps a system of physical relations).

There is a further problem understanding the relation between *W* and a spacetime region at this preliminary stage of the exposition, insofar as

> The explicit construction of *W* is the main objective in this book. (51f).

The picture painted in Rovelli and Vidotto (2015, section 2.4.3) is not of processes undergone by systems during an interval of time while they evolve undisturbed, punctuated by episodes during which they interact with one another. The notion of a background time has disappeared along with that of a background spacetime. That picture has been replaced in LQG by one of systems as processes *constituting* spacetime regions, and their interactions as occurring at the boundaries between these regions.

Perhaps these problems arise because we have been taking this picture too seriously. In LQG, (classical) spacetime regions are supposed to emerge only in a suitable limit of the theory, while at the fundamental level

> as far as we know today, all that exist in nature are general-covariant quantum fields. (19)

Until the construction of *W* is successfully completed, it is impossible to say just what a spacetime region is, making it problematic even to talk of spacetime regions while addressing the measurement problem in LQG. Now, if spacetime is to emerge only in some limit of the theory, one may not expect to be able to talk significantly about spacetime regions away from that limit. So perhaps charity would recommend that the discussion of Rovelli's relational interpretation in section 2.4.3 of this book be taken metaphorically rather than literally. But doing so would leave it unclear how to apply Rovelli's relational interpretation in this context.

Rovelli and Vidotto (2015) give the physical theory of quantum gravity only in chapter 7, where its relational interpretation is not discussed. But I think a brief examination of that theory helps resolve some of the unclarity of the earlier discussion in section 2.4.3. The final expression for the transition amplitude *W* given there (by equations (7.46), (7.47) on page 141) is stated to give the theory of LQG *on a given 2-complex* (Rovelli and Vidotto 2015, 142).

A 2-complex is neither a region of spacetime nor a collection of such regions. It is the dual of a bulk triangulation Δ of a *mathematical* four-dimensional space with the topology of (a compact region of) $\mathbb{R}^4$. It is true that Δ is introduced (on page 131) as a triangulation of a compact region *M* of spacetime. But it is common practice for theoretical physicists to elide the distinction between (regions of) physical spacetime and the elements of a mathematical object used to represent them. And to take *M* to represent a region of physical spacetime would conflict with LQG's overarching goal of retrieving spacetime only in an appropriate limit of the theory.

Indeed, Rovelli and Vidotto (2015, section 7.2.3) warn their reader against taking the geometrical picture accompanying their mathematics too literally:

> ... the geometrical picture of 'tetrahedra', 'triangles', etc., must be taken as something meaningful only in some classical approximation and not at the fundamental scale.
>
> There are quantum states, formed by quanta of the gravitational field, which have the property of giving rise to something that we describe as a three-dimensional Riemannian geometry in the limit of large quantum numbers. (140)

They even warn against a naive physical interpretation of such talk:

> ... the quanta of space of loop quantum gravity should not be taken too naively as actual entities, but rather as modes of interaction. (141)

The picture painted in Rovelli and Vidotto (2015, section 2.4.3) is not of physical processes in spacetime, but of elements of mathematical structure basic to the theory of LQG that eventually predicts transition amplitudes for physical processes that have yet to be interpreted. Interpreting LQG *is* interpreting these processes. Nothing that is said in Rovelli and Vidotto (2015, section 2.4.3) contributes directly to that project. But this discussion may yet contribute *indirectly* to a resolution of the measurement problem in LQG.

10.4 The Measurement Problem

Leggett (2005) stated the quantum measurement problem this way:

> most interpretations of quantum mechanics at the microscopic level do not allow definite outcomes to be realized, whereas at the level of our human consciousness it seems a matter of direct experience that such outcomes occur. (871)

Rovelli responds by taking a measurement interaction between *S* and *O* to have an outcome relative to *O*, but not relative to an external *O′* before it interacts with them, as in this picture by Rovelli (2004) in Fig. 10.1.

This depicts *O* and *O′* as *human* observer systems, despite Rovelli's insistence that observers are not physically special systems in any sense. But while *O* must be a physically special system to be capable of conscious human *experience* of the outcome relative to *O*, Rovelli can consistently claim that *there is* such an outcome relative to *O* whether or not *O* is physically special in this way. In

Fig. 10.1 The observer observed

his view, an outcome relative to an observer need not be experienced by that observer. Insofar as typical observer systems are not conscious, most outcomes of measurement interactions are not experienced.

What should we say about O' prior to any interaction with S or O? Certainly O' can not then consciously experience an outcome, whether or not it has the capacity for conscious human experience. Rovelli goes further by denying that *there is* then any outcome relative to O': for there to be any outcome relative to O', O' must first interact appropriately with S or O. So before O' interacts with S or O, there is no outcome relative to O': there is only an outcome relative to O.

Despite the deliverances of conscious human experience, quantum measurements have no objective, observer-independent outcomes in Rovelli's relational quantum mechanics.

10.5 Relational Loop Quantum Gravity

Rovelli's exposition of relational quantum mechanics (Rovelli 1996) leans heavily on the model of simple von Neumann interactions between quantum systems such as S, O that correlate an initial pure eigenstate $|q_i\rangle_S$ of $\hat{q}$ on S with a corresponding final pure eigenstate on O.

$$|q_i\rangle_S \otimes |ready\rangle_O \Longrightarrow |q_i\rangle_S \otimes |\text{“}q = q_i\text{”}\rangle_O \tag{1}$$

While 'observer system' O need not be physically special for this interaction to yield a value q_i of the quantity q on S relative to O (which may even be an electron), only if O is a special physical system can O be said to have knowledge of q as a result of this interaction. Human scientists are special physical systems that may be idealized as quantum systems: perhaps registration by a non-conscious measuring apparatus or another kind of IGUS (Information-Gathering and Utilizing System) could also be said to constitute knowledge.

How can this model be transposed into the context of LQG? The arrow $\Longrightarrow$ of equation (1) cannot be replicated at a fundamental level, since there is no time at a fundamental level in LQG. Nor is it clear how to replicate the systems represented in equation (1) if all that exist in nature are general covariant fields.

Suppose we were to take the picture of Rovelli and Vidotto (2015, section 2.4.3) to provide the basis for a fundamental LQG model of a measurement interaction. In that picture an interaction is what happens at the boundary between two spacetime regions. These regions are now the systems that interact, conceived as processes rather than enduring objects. But how are we to model an observer system capable of registering if not experiencing the (relative) outcome of the interaction?

An observer system O may be reasonably idealized in quantum mechanics as an enduring quantum system composed of a vast number of particles with bulk degrees of freedom corresponding to recording states. No analogous idealization is apparent in LQG: an observer system is not reasonably idealized as a spacetime region! But eventually, fermion and boson quantum fields must be integrated into LQG. So one might hope to use these to construct an idealized model of an observer system in LQG. However, this remains a distant aspiration. There is no present prospect of a satisfactory model of a measurement interaction at the fundamental level in LQG.

But we saw in section 10.3.1 that the discussion of Rovelli and Vidotto (2015, section 2.4.3) is best taken to describe a geometrical picture not of the structure of (processes in) spacetime in LQG, but of the mathematics of transition amplitudes defined over bulk triangulations and their duals in $\mathbb{R}^4$. It is a category mistake to identify these mathematical structures on $\mathbb{R}^4$ with regions of physical spacetime. LQG does not itself *describe* regions of spacetime, either classical or nonclassical. It provides (picturable) mathematical models that may be used to calculate transition amplitudes. The problem is to understand what these are transition amplitudes *for* in the absence of objects or processes in physical spacetime.

Now it is a constraint on the empirical adequacy of LQG that classical spacetime emerges in a suitable limit of the theory. Assuming that constraint is met, physical spacetime may be said to exist in that limit, along with whatever physical processes occur within it. At a fundamental level, those processes may be modeled by theories of fermion and boson quantum fields as well as the LQG field. It is a constraint on the empirical adequacy of those other field theories that ordinary quantum mechanics emerge in a suitable limit, just as it is a constraint on the empirical adequacy of LQG that gravitational processes in emergent classical spacetime be correctly modeled. So while the von Neumann model (1) cannot be replicated at the fundamental level in LQG, it must be recoverable from that theory in a suitable limit in which systems like *S*, *O* emerge as objects that endure in classical spacetime. In that limit, the transition amplitudes for gravitational processes calculated from LQG may be applied in models like (1) since such a unitary transformation of states is calculable from them.

While LQG cannot model von Neumann interactions at a fundamental level, its transition amplitudes must be compatible with the existence of such interactions in a suitable limit in which classical spacetime emerges, along with enduring physical systems to play the roles of system *S* and observer *O*. That is how LQG may be given a relational interpretation along the lines of Rovelli (1996). But does this help solve the measurement problem for LQG?

10.6 The Measurement Problem in Relational Loop Quantum Gravity

There can be no *absolute* definite outcomes in relational LQG, since all states are relative to systems. The measurement problem is then to reconcile merely *relative* outcomes with the fact that at the level of our human consciousness it seems a matter of direct experience that definite outcomes occur. In relational quantum mechanics, this may be possible if one models a human observer as a (special kind of) enduring physical system.

Even if successful, this strategy comes at a steep cost, because it sacrifices an ideal of scientific objectivity: it has always been a basic scientific norm that a human observer's sincere report states what has objectively happened. This norm is violated if *O*'s sincere report of the outcome of his quantum measurement merely reports its outcome *relative to him*, while for *P* the measurement then *has* no outcome, but only acquires an outcome *relative to her* after she has (directly or indirectly) interacted with *O* or *S*.

Perhaps this norm is not inviolable. Perhaps science could continue without it as long as no further observations, by *O*, *P* or any other observer, could make this violation manifest? Rovelli (1997) has appealed to quantum models of measurement-type interactions to show why it must remain hidden: in these models, repeated careful measurements by any observer, on the original system or on other observers, will yield *relative* outcomes in *apparent* conformity to that norm.

The absence of an LQG model of a human observer presents an obstacle to implementing such a relational solution to the measurement problem in LQG. A human observer is a very special physical system for which we have no model in LQG: there is no present prospect of modeling even a simple recording apparatus in LQG. But as we saw in section 10.5, it is not necessary to model a human observer directly in LQG in order to implement a relational interpretation of that theory. Within an empirically adequate LQG, a model of a measurement interaction in ordinary quantum mechanics in a classical background spacetime will provide a platform on which to build such a relational interpretation. Whether such a platform is available hinges on the successful treatment of the limiting process in LQG.

10.7 Morals of the Story

We have learned two morals from this exploration of Rovelli's strategy for solving the measurement problem in LQG.

1. Relativizing values of magnitudes as well as quantum states to systems (and only after they have interacted) not only abandons the notion of a true, universal, observer-independent description of the state of the world, but also threatens a basic norm of scientific objectivity.
2. In addressing the quantum measurement problem, it is neither necessary nor possible to model an observer system at a fundamental level within LQG. A physical characterization in terms of structures (including classical spacetime) that emerge from LQG only in an appropriate limit is all that is needed. So a resolution to the measurement problem is not a prerequisite for the recovery of spacetime in quantum gravity.[2]

Taking these morals to heart suggests certain modifications to Rovelli's relational strategy:

By allowing relativization of quantum states *but not values of magnitudes* to physical systems, it may be possible to retain this basic norm of scientific objectivity, and even the notion of a true, universal, observer-independent description of the state of the world. This means rejecting the eigenvalue-eigenstate link. It is not necessary to treat all physical systems on a par when relativizing *quantum* states to physical systems. Relativization to physically special observer systems will suffice. A quantum theory of gravity is not itself required to say what makes these observer systems special in order effectively to address the measurement problem:

[2] But of course an adequate understanding of *any* quantum theory does require that this problem be addressed and solved or dissolved at some stage. Note that in drawing this moral I register disagreement with Wüthrich (2017, 332) while gratefully acknowledging the stimulus his paper and subsequent comments have provided to my own thinking.

that characterization may be given in terms of physical structures (including spacetime) that emerge only in a suitable limit of the theory.

A pragmatist approach to quantum theory (Healey 2012, 2017) takes these morals to heart when applied to the emergence of spacetime in a prospective quantum theory of gravity (Healey 2020).

References

Fuchs, C., N. D. Mermin, and R. Schack (2014). An introduction to QBism with an application to the locality of quantum mechanics. *American Journal of Physics 82*(8), 749–754.

Fuchs, C. A. (2010). QBism, the perimeter of quantum Bayesianism. arXiv 1003.5209.

Healey, R. (2012). Quantum theory: A pragmatist approach. *British Journal for the Philosophy of Science 63*(4), 729–771.

Healey, R. (2017). *The Quantum Revolution in Philosophy*. Oxford: Oxford University Press.

Healey, R. (2020). On the independent emergence of spacetime. In D. Glick, G. Darby, and A. Marmodoro (Eds.), *The Foundation of Reality: Fundamentality, Space and Time*, pp. 183–194. Oxford: Oxford University Press.

Leggett, A. (2005). The quantum measurement problem. *Science 307*(5711), 871–872.

Rovelli, C. (1996). Relational quantum mechanics. *International Journal of Theoretical Physics 35*(8), 1637–1678.

Rovelli, C. (1997). Relational quantum mechanics. arXiv quant-ph/9609002v2.

Rovelli, C. (2004). *Quantum Gravity*. Cambridge: Cambridge University Press.

Rovelli, C. and F. Vidotto (2015). *Covariant Loop Quantum Gravity*. Cambridge: Cambridge University Press.

Wüthrich, C. (2017). Raiders of the lost spacetime. In D. Lehmkuhl, G. Schiemann, and E. Scholz (Eds.), *Towards a Theory of Spacetime Theories*, pp. 297–335. Basel: Birkhäuser.

11

The 'Philosopher's Stone'

Physics, Metaphysics, and the Value of a Final Theory

Kerry McKenzie

Abstract

Physicists have long been in search of the final theory—a physical theory that can be regarded as the truly fundamental description of nature. But metaphysicians likewise aspire to describe the world as it is most fundamentally. I argue that if we take a naturalistic approach to metaphysics, a final theory is even more crucial to success of the metaphysical project than it is to that of the physicist. This is because the non-fundamental theories produced by contemporary physicists may at least be said to approximate the final theory, and so physicists may be said to at least be making progress towards their goal in advance of having achieved it. Metaphysical theories, by contrast, cannot be said to be 'approximately true', and hence do not obviously partake in such progress. This raises questions of the value of engaging in naturalistic metaphysics prior to the emergence of a truly final theory.

11.1 Introduction

For all the profound advances that have been made in modern physics, theoretical physicists are still 'dreaming of a final theory'—a theory of physics that can at last be regarded as a truly fundamental description of nature.[1] Since all the 'arrows of explanation' seem to point at this theory being a quantum theory of gravity, it is this that physicists have variously identified with 'the theory of everything', 'the philosopher's stone', and 'the ultimate triumph of human reason'.[2] While the more extreme of these statements might at times make us wince, it can hardly be denied that the discovery of such a theory would be an epistemic achievement of great significance. However, it is not only fundamental physicists that long for the final theory: for metaphysicians, at least of a naturalistic stripe, await the emergence of

[1] Weinberg (1994). [2] Weinberg (1994); Hawking (1998, 193); Finkelstein (2006, 25).

Kerry McKenzie, *The 'Philosopher's Stone': Physics, Metaphysics, and the Value of a Final Theory* In: *Philosophy Beyond Spacetime: Implications from Quantum Gravity.* Edited by: Christian Wüthrich, Baptiste Le Bihan, and Nick Huggett, Oxford University Press. DOI: 10.1093/oso/9780198844143.003.0011

this theory too. For metaphysics is often conceived of as principally concerned with articulating the fundamental, even if different aspects than the physicists typically focus on; indeed, some have gone as far as to say that metaphysics concerns the fundamental alone.[3] And since 'naturalistic' metaphysics—that to which metaphysicians of physics typically commit—is by definition metaphysics 'informed' by the relevant science, it is only with the guidance of a physical theory of the fundamental that such a project can proceed. It thus seems that many metaphysicians of physics, whether they realize it or not, are 'dreaming of a final theory' also—for it seems a *sine qua non* of a naturalistic metaphysics of the fundamental.

Fundamental physicists and naturalistic metaphysicians, then, would seem to hold much in common—sharing as they do in their aspiration to describe the fundamental and in their need of a final physics theory to do so. But what I want to argue here is that the metaphysicians are even more greatly in need of that theory than the physicists are. They are even more in need because—insofar as it is their aim to describe the fundamental—it is entirely unclear that they can enjoy *any measure of success whatsoever* towards this end until such time as that theory is in hand. For it is not at all clear that fundamental metaphysicians can make claims that possess *any measure of truth* so long as we are in lieu of the final theory, or that they can claim to be *progressing towards* doing so. By contrast, today's physicists it seems can make claims that can be regarded as being at least 'approximately true' of the fundamental—or, at the very least, as 'closer to the truth' than were the claims that they made before. This means that, unlike in the case of metaphysics, there is an intuitive sense in which we may regard contemporary physicists as making progress towards their goal of describing the fundamental, even if they are still far from reaching it. At the root of this contrast—or so I shall argue—is the fact that the very concept of 'approximation', while routinely predicated of claims of physics, seems to make little sense in application to paradigmatic claims of metaphysics. As a result, the notion of such claims 'getting closer to the truth' would seem to make little sense in that context too. And since, as we will see, metaphysical claims about the fundamental based on our current non-fundamental theories are unlikely to be true *simpliciter*, what the value is of engaging in fundamental metaphysics before the 'end of science' reveals itself to be very unclear.

That a final theory is even more indispensible to the ends of metaphysicians than it is to fundamental physicists is in any case what I want to venture here. Before I make the argument, however, let me make two clarifications on its scope and limitations. Firstly, it is not another wholly generic argument against the viability of metaphysics, such as those based on 'metaphysical underdetermination' and so on; rather, it is directed at those already harbouring metaphysical sympathies,

[3] See, among many others, Sider (2001, 1); Lowe (2001, 2).

and attacks *naturalistic fundamentality metaphysics* in particular. As such, it has nothing to say to those who—albeit for reasons perennially puzzling to me—believe that the fundamental nature of reality can be intuited without the guidance of science. Nor does it (or at least not directly) have a bone to pick with those metaphysicians focused on the non-fundamental, such as those defending the existence of 'effective' physical ontology (although it may place limits on how distinctively metaphysical is the vocabulary we can employ in its description).[4] But as I hope to convey in the next section, a great deal of work in the metaphysics of science could rightly be regarded as 'naturalistic fundamentality metaphysics', and so of the sort to be subsumed by my argument. Secondly, although the argument to be put forward here of course strikes a rather sceptical note with regard to the value of (much) metaphysics, it may be more productive to *modus tollens* my *modus ponens* and view it as an invitation to reflect on what progress *is* made in philosophy and in metaphysics in particular, and hence on the value of what we metaphysicians of science do. After all, given that philosophers of science more generally have not taken progress in science for granted, but instead have wrestled with questions of what such progress might consist in and whether science may really be said to enjoy it, it seems only appropriate for metaphysicians of science to engage in analogous reflections regarding their own line of work. Furthermore, given how scathing many naturalistic metaphysicians have been of the value of armchair metaphysics, based as it is on at best outdated science, it seems borderline hypocritical to blithely assume that there is value in what *we* do, insofar as we know that we routinely speculate on metaphysics of the fundamental while still bereft of the theory that seems so obviously needed in order for us to meaningfully do so.

The structure is as follows. In Section 11.2, I clarify what is meant by 'naturalistic fundamentality metaphysics', and underline that many of the central debates in metaphysics of physics can be categorized in this way. In Section 11.3, I remind us of some of the ways in which claims of metaphysics have been thought to change in truth-value as science progresses—raising the question of whether we can be confident in the metaphysical claims that we make today. In Section 11.4, I consider how, even given analogous facts of theory-change, our confidence in the case of *scientific* claims can arguably be retained, on account of an available notion of 'approximate truth'. In Section 11.5, I argue that such a defence is unavailable for canonical claims in the metaphysics of physics, at least as ordinarily understood. Section 11.6 is a brief conclusion.[5]

[4] See e.g. Williams (2019). (The 'limitations' alluded to here will be unpacked in section 11.5 below.)

[5] The historically motivated problem for metaphysics that I will outline here has also been broached by Bradley Monton (2011). The current chapter attempts to go beyond that work, however, in that it attempts a diagnosis of it.

11.2 What Is 'Naturalistic Fundamentality Metaphysics'?

First on the agenda is demarcating more carefully what it is that is meant by 'naturalistic fundamentality metaphysics'—the target of the present critique. 'Fundamentality' metaphysics is, as the name suggests, metaphysics primarily concerned with that which is fundamental—typically, although not always, together with articulating the relation it bears to the non-fundamental. In its contemporary incarnation, this is often conceived as the conjunction of two distinct projects, modelled in terms first put forward by Quine: that of identifying the *fundamental ontology*, and that of identifying the *fundamental ideology*.[6] Typically (though not always), the fundamental ontology at issue in these debates is the fundamental *categories* of the world. The 'fundamental ideology', by contrast, is the suite of predicates or operators that are taken as primitive in the theory, and hence in terms of which all other concepts of the theory are to be defined. As such, they are the predicates and operators which appear in the most *parsimonious expression* of the theory. Since in either case what fundamentalist metaphysics does is analyse ontology or concepts into (what we could call) 'lowest terms', whether the basic categories or the basic concepts, we may regard fundamentalist metaphysics as 'analytic' metaphysics in the literal sense. However, since that term has come to connote armchair or *a priori* metaphysics—i.e. precisely that with which naturalistic metaphysics is contrasted—I will not use that term here. 'Naturalistic' metaphysics, on the other hand, is (very roughly) metaphysics that uses the relevant science to 'inform' both the metaphysical questions we deem it appropriate to ask and the answers that we give to them. Here we make no pretence to having defined it; but hopefully we can just agree that we know it when we see it. And I think it is clear that much of what we are inclined to call debates of naturalistic metaphysics are also debates in fundamentality metaphysics.

Consider, for example, the debate over the tenability of ontic structural realism (OSR). Claimed by its protagonists as the exemplar of naturalistic metaphysics *par excellence*, it is explicitly a thesis of fundamentality metaphysics—namely, the thesis that "[objecthood] is to be understood as derivative at best, with structure as the fundamental ontological category".[7] This denial of the fundamentality of objects may be understood as the denial of the fundamentality of *pluralities* of entities with *intrinsic natures*. The plurality requirement on objects is there to render objects *particulars*, as distinct from *universals* (the latter being conceived of as 'a one over a many'); objects, then, must be the sort of things that can meaningfully admit of qualitative *duplication*. This plurality requirement of course entails that, when we are speaking of objects, we are speaking of entities about

[6] Quine (1951, 14). [7] French (2012, 122).

which we can say *how many there are*, and as such ascribe a *cardinality*.[8] And the conception of objects as entities with 'intrinsic natures' has been embedded in structuralism from the very beginning, with Grover Maxwell identifying structural features with 'those that are not intrinsic'—namely, relations and extrinsic properties.[9] Since the properties of physical objects are taken to resolve into their qualitative properties (such as the charge and mass definitive of the various particle types) and their identity-conferring properties (which differentiate tokens of the same type), the denial of 'intrinsic natures' amounts to the denial that *any* of these properties are intrinsic.[10] Thus in denying the fundamentality of the category of objects structuralists hold that at the fundamental level there either are no pluralities of entities with intrinsic natures (this defining the eliminativist version of structuralism); or that, if there are, they can nevertheless be accounted for wholly in structural terms (this defining the priority-based version).[11] Now, as a debate over which category is fundamental, the debate over OSR is most naturally construed as a debate over *primitive ontology*. However, identity considerations also being so central to structuralism, it may also involve arguments as to the correct *primitive ideology*—namely, arguments over whether the identity predicate featuring in physical theories may be defined in structural terms.

Next consider perhaps the most central debate in the metaphysics of science: the debate between Humeans and anti-Humeans. While what Hume would have intended by 'Humeanism' is a matter of dispute (and his reasons for being 'Humean' out of fashion), Humeanism in its contemporary guise is the thesis that "modality is to be analyzed, or at least accounted for, in non-modal terms".[12] As such, the goal of contemporary Humeans is that of "accounting for modality in a fundamentally amodal world".[13] Hence the dispute between Humeans and anti-Humeans is not over whether modal claims such as 'I could have been a contender' are true or false; or whether it is true that smoking causes cancer; or whether Fresnel's law is a law; or whether any other claim involving modal paraphernalia we can think of has a truth-value. Rather, it is a dispute over whether what *makes* such claims true or false can ultimately be specified without recourse to modal concepts. Thus, concerning as it does how such concepts and categories are analysed, this debate is most naturally construed as a debate over the fundamentality of modal ideology such as $\Diamond$ and $\Box$ operators. David Lewis' system of possible world semantics is widely taken as the only system that can facilitate the comprehensive analysis of modal language required for Humeanism to go through. As such,

[8] French and Krause (2006), for example, do not insist that objects be regardable as individuals, only that a cardinality can be ascribed to them.

[9] Maxwell (1970). See McKenzie (2020) for further discussion.

[10] Chakravartty (2012, 204).

[11] See again McKenzie (2020).

[12] Wilson (2015, 147).

[13] Sider (2013, 270).

most contemporary Humeans model their theories on his system in some way or other.

For a final example, consider the debate in the philosophy of time over the doctrine of *presentism*. Although it comes in various versions, presentism is at heart the *prima facie* intuitive thesis that all that really exists is that which exists *now*. Here it contrasts with its rival *eternalism*, which holds that every event—including those we call 'past', 'present', and 'future'—enjoys an eternal existence. Despite its intuitive appeal, a problem for presentism is that it has trouble accounting for such uncontroversial truths as 'There were dinosaurs', given that according to it there is, at present, no truthmaker to make true a token of this sentence uttered now. As such, it is now understood that commitment to presentism inevitably involves a commitment to a certain ideological package—namely, primitive WILL and WAS operators, whose function is to map present-tensed sentences to truth-values.[14] Since eternalism by contrast can provide truthmakers for such claims—such as the fact that there is (that is, there is *now*) a past event featuring rampaging dinosaurs—no such temporal ideology is compulsory in this theory. The ideological parsimony advantages thereby on offer to eternalism are for some taken as an important point in its favour.[15] More impressive to naturalists, however, is the fact that its main rival of presentism faces formidable challenges in the face of relativity, making eternalism by now a paradigm doctrine of naturalistic metaphysics.[16] But what that means is that this paradigm naturalistic debate is also a debate over the correct fundamental ideology.

It seems, then, that some of the central debates in (what we are apt to call) naturalistic metaphysics are debates in fundamentality metaphysics: debates over what is ontologically and what is ideologically fundamental. As already flagged up in the introduction, however, it is on reflection somewhat odd that naturalistic metaphysicians are so preoccupied with debating what is fundamental, given that we do not have the final theory that we think is needed in order to 'inform' their claims. For while naturalism is often supported by the fact that scientific developments have fatally undermined *a priori* speculations about the nature of reality—the necessity of determinism or the Euclidean character of space being just two examples—science has also repeatedly undermined metaphysical claims that have been motivated by previous science. And since—given that we lack a fundamental theory—we think that our present theories too will in time be superseded, this suggests that *they* cannot be regarded as indicative of the fundamental metaphysics either. Since this point lies at the core of the present argument, it is worth elaborating on in some detail.

[14] See Sider (2013, section 11.2). [15] See e.g. Sider (2001). [16] See e.g. Putnam (1967).

11.3 Theory Change in Metaphysics

Take again our canonical examples of debates of naturalistic fundamentality metaphysics: the debate over OSR versus 'object-oriented' realism; over Humeanism versus anti-Humeanism; and over presentism versus eternalism. We have seen that these are, at least in some essential part, naturally construed as debates over what is fundamental, whether in terms of ontology or ideology. As such, we may list some of the canonical claims of naturalistic fundamentality metaphysics as answers to the following questions:

Questions of Primitive Ontology:

1. Which *categories* are fundamental? Are *objects* fundamental? That is, are there fundamentally entities:
   1a. to which some *intrinsic* properties are essential?
   1b. to which we can ascribe a *cardinality*?

Questions of Primitive Ideology:

2a. Do objects need *primitive identity*, or are identity facts made true by facts stateable wholly in qualitative terms?
2b. Do we need *primitive modality*, or are modal truths made true by facts stateable wholly in non-modal terms?
2c. Do we need *primitive tense operators*, or are tensed truths made true by facts stateable in wholly tenseless terms?

These, as has been pressed, are some of the paradigmatic questions of naturalistic metaphysics; since they concern the fundamental, they are also questions of fundamentality metaphysics. The problem, however, is that the answers to them have all been argued to change as we have transitioned from theories once regarded as fundamental to theories now regarded as more fundamental. As Monton (2011) discusses how the viability of presentism may be sensitive to changes in physical theory, here I will focus on the remaining questions, working backwards through the list.[17]

2b. Do we need primitive modality, or are modal truths made true by facts stateable wholly in non-modal terms? It has already been mentioned that Lewis' system is the only system understood to have the potential for furnishing a comprehensive reductive analysis for modal language. And as Lewis himself was perfectly open about, his system was 'inspired by classical physics' (Lewis, 1998,

[17] See Wüthrich (2010) for an in-depth discussion of the prospects of presentism in quantum gravity.

222). However, the idea his system can survive the transition to quantum physics has been put under repeated pressure. To present the problem in just one way, note that in order for Lewis' system to succeed in its reductive ambitions, it must be able to guarantee that for every way the world could be there is a world that is that way—the so-called principle of plenitude. And the way that Lewis attempts to secure this is by assuming that the fundamental properties respect the *principle of free recombination*—the principle that the spatiotemporal instantiations of such properties may be rearranged in wholly arbitrary ways. This principle carries certain implications as to the nature of fundamental properties; most saliently, that they are *intrinsic* and *categorical*. For only in that case will these properties be such that 'their instantiation has no metaphysical implications concerning the instantiations of fundamental properties elsewhere and elsewhen'.[18] Now, since this process of rearrangement is that which, in Lewis' system, generates new possible worlds from old, it is this that generates the fundamental truthmakers for modal claims, and it is the wholly *unrestricted* nature of this recombination that is taken to underwrite the idea that there are no 'gaps' in possibility space, in accordance with plenitude. However, the idea that spatially separated instantiations of properties are independent of one another in the way the principle of recombination requires is widely regarded as incompatible with quantum mechanics, because of the phenomenon of entanglement. For when we have entangled systems we find that fundamental properties defined on spatiotemporally separated regions are not independent but rather coordinated with each other, such that it is possible to forecast with certainty results of measurements carried out on one region given information about measurements carried out on another remote from the first. As such, if the fundamental properties of the world have the metaphysical profile apparently attributed to them by quantum mechanics, it seems that their instantiations do not have the features required of them by the principle that lies at the heart of the supposed success of Lewis' modal analysis. And since the absence of such remote coordination is arguably 'the most natural way for the Humean to guarantee that the world be fundamentally non-modal and lacking in necessary connections', the prospects for some other reductive analysis succeeding would at least appear to be greatly compromised.[19]

Of course, by now a huge amount of ink has been spilled on whether the 'spirit' of Humeanism is consistent with quantum entanglement after all. Ismael and Schaffer (2016), for example, have defended the view that in quantum contexts it is not the mosaic of 'local matters of fact' that ought to be regarded as fundamental, but rather the cosmos as a whole. This would give a common ground to the

[18] Loewer (1996, 177). [19] Bhogal and Perry (2017, 91).

non-local correlations, rendering them no longer necessary connections between *distinct* existences—the *bête noire* of Humeanism. For this project to succeed, it would need to be shown that worlds so characterized may be arranged with respect to a metric that will ensure that there are sufficiently many of them to ensure a truthmaker for every true modal claim—something not entirely obvious given that worlds are now characterized in such monolithic terms. But in any case, the embrace of holism as a metaphysical response to preserve Humeanism arguably represents a different but equally radical departure from the received classical metaphysics. An alternative solution, however, is to abandon the view that ordinary 3-space is fundamental, and supplant it with configuration space—a vastly enlarged space of 3*N* dimensions, with *N* the number of particles involved.[20] For it may be shown that no correlations between distinct points of *that* are needed to account for the distant correlations in 3-space. However, that approach runs into problems when we make quantum theory relativistic—for in that context particle number is not conserved, or in general well defined, making configuration space ill defined also.[21] And again, such a move would seem to concede that which is relevant to our purposes—namely, that the metaphysics appropriate to the quantum domain is radically different from that which is natural to the classical (space no longer being viewed as fundamental, for example). Wherever it is precisely that the bump in the carpet goes up, then, it seems *something* in the metaphysics of qualitative properties, their relationship to space and time, and the status of modal discourse is going to have to change radically under the transition from classical to quantum mechanics.[22]

2a. Do we need primitive identity, or are identity facts made true by facts stateable wholly in qualitative terms? This question has been principally pursued in the context of structuralist metaphysics. As such, the focus here is normally on whether facts concerning the *identities of objects* can be reduced to qualitative facts (for structuralists, facts involving only structural features), as opposed to having to be taken as unanalysable and hence primitive. This question is standardly understood to fraction into the questions of the *synchronic* identities of objects (in virtue of which an object is the object that it is and not another temporally co-existent) and their *trans-temporal* identities (in virtue of which an object counts as the same object over time). Although this topic has generated a huge amount of discussion—which is of course code for disagreement—I think it is fair to say that almost everyone agrees that some aspect of this question must undergo change

[20] See e.g. Loewer (1996).

[21] See e.g. Myrvold (2015)—although he stresses that there may be moves, however unnatural, that the defender of configuration-space realism may make.

[22] As Peter Lewis puts it, "very little can be concluded unconditionally on the basis of quantum mechanics . . . The best we can say is that not everything in our received classical worldview can be right" (2016, 187).

from classical to quantum mechanics. For the answer in classical mechanics seems to be an unambiguous 'no' in both cases; however, seemingly everyone agrees that no notion of trans-temporal identity may be defined in the quantum context (except in trivial cases).[23] For even if two otherwise indistinguishable quantum particles begin in distinct position eigenstates, the corresponding wavefunctions will in general quickly spread until such point as they overlap, meaning that points in a sequence of position measurements may not be unambiguously attributed to any one particle over another (even if it is vastly more probable for some one).[24] If we demand trans-temporal identities for objects, then, it seems that they must be taken as primitive.[25] Here again, then, the classical-quantum transition brings in its wake deep changes in our fundamentality metaphysics.

1b. Are there fundamentally entities to which we can ascribe a cardinality? This is a question raised in connection with considering the fundamentality of objects (and hence the status of ontic structuralism), and the paradigm notion of a fundamental object in physics is surely the 'elementary particle'. And the answer to whether we can ascribe cardinalities to particles is taken to change as we transition between quantum theories formulated on flat and on general curved spacetimes. For it may be shown in quantum field theory (QFT) that as the curvature tensor becomes non-zero, the number of particles detected in a region becomes *observer-dependent*—a phenomenon known as the 'Unruh effect'.[26] Given that—at least in general curved spacetimes—all observers are regarded on an equal footing physically, this has been argued to put pressure on the *objectivity* of the particles to which a cardinality is attached.[27] As Baker puts it, "Since one observer's particle concept may ascribe non-zero particle number to the state another observer would call a zero-particle vacuum, there is *no objective fact* about the particle content of QFT states in curved spacetime".[28] As such, since what is *ontologically fundamental* is presumably also *real and objective*, it seems the answer to this question changes

[23] Regarding the (more ambiguous) synchronic case, see the discussion following the publication of Saunders (2003).

[24] Dalla Chiara and Di Francia (1993, 266); Huggett (1997).

[25] Of course, in the face of this one might well simply deny that quantum objects do in fact *have* any trans-temporal identity, and thus deny that there is such any property to have to be taken as primitive. But that again would support radical changes with respect to our primitive ontology. For this is standardly taken to imply that they are no longer *objects* at all, at least understood as a category distinct from *events*. (Schrödinger was an early proponent of this view; see Bitbol 2007, 88.)

[26] Wald (1994, 116); Crispino et al. (2008).

[27] I cannot here go into the chain of argument that has led physicists to the view that "what is invariant in a class of admissable coordinate systems represents the objective contribution of reality", as opposed to the "subjective contribution" that we must make in order to get physics off the ground (Friedman 1983, 19). But see Brading and Castellani (2003) for extensive discussion.

[28] Baker (2013, 269), italics added. Similarly Wald (1994, 116): "The Unruh effect may appear paradoxical to readers who are used to thinking that quantum field theory is, fundamentally, a theory of 'particles', and that the notion of 'particles' has objective significance".

as we transition between theories formulated in flat and curved spacetimes—hence, we think, from less to more fundamental quantum-theoretic descriptions.

1a. Are there fundamentally entities to which some intrinsic properties are essential? The answer typically given to this question is 'yes', for it is standardly assumed that the fundamental kind properties—the properties of determinate mass, charge, spin, etc., definitive of the fundamental kinds—are intrinsic properties. However, while this may be the case for such properties as understood in classical physics, I have argued elsewhere that in the context of *QFT* the picture is radically changed. For it seems that it can no longer be maintained in fundamental—that is, high-energy—quantum field-theoretic regimes that any of the kind properties may be regarded as intrinsic, at least in the familiar sense of 'independent of accompaniment'.[29] The reason is that fundamental fields occur in quantum theories which are consistent in the limit of unlimited energy (equivalently, in the limit that the distances described by the theory get arbitrarily small). But it turns out that only theories containing very special combinations of fields have this property of being well defined in the $E \rightarrow \infty$ limit. As such, arbitrarily adding and subtracting fundamental fields from worlds is more apt to produce mathematical nonsense than a coherent description. But then it seems that the possession of properties by fundamental fields cannot be regarded as 'independent of accompaniment' after all: on the contrary, fine-tuning of field content emerges as necessary for there to be a world containing fundamental fields at all.[30]

It seems, then, that it is quite routine for the most natural answers to questions of naturalistic fundamentality metaphysics to undergo change as a given physical theory is abandoned as a fundamental description and supplanted by a more fundamental successor. Again, this in one sense should in no way surprise us, since the prime motivation for naturalistic metaphysics is precisely the fact that the plausability of claims of metaphysics has proved sensitive to developments in science. But since we also know that our *pro tem* fundamental theories are likely not truly fundamental, we have inductive reason to expect our current fundamentality metaphysics to be abandoned in its turn as well. And since it is unclear what the *purpose* of metaphysics is other than to create theories that we can regard as true, why we should bother speculating about the fundamental while we await the final theory becomes distinctly unclear too.

These considerations make it seem as if metaphysical enquiry into the fundamental must be postponed indefinitely. However, one might think that there is an obvious rejoinder to this axiological challenge to metaphysics—moreover, one thoroughly naturalistic in spirit. For on the face of it, it seems the same sceptical

[29] Langton and Lewis (1998). We will have more to say about the definition of intrinsicality below, but almost everyone takes this criterion as at least necessary for intrinsicality (cf. Figdor 2014).

[30] See McKenzie (2016) for more detail.

argument from theory change just applied to fundamentality metaphysics could be applied equally to fundamental physics. For that enterprise also aims to describe the fundamental, and as yet lacks the resources needed to do so; yet clearly scientists and philosophers of science think that there is value in physics! However, as I will argue in closing, that which lends value to contemporary physics *qua* the project of describing the fundamental does not apply to fundamental metaphysics. As such, there is no easy inference from the idea there is value in the former to the idea that there is value in the latter. To see this, let us recap how it is that the problem of scientific theory change has been dealt with in the philosophy of science. While that story might encourage the thought that the metaphysics of physics can also circumvent the problem, we will then see, in Section 11.5, how the defensive moves that may be invoked to salvage science cannot be made in the case of its attendant metaphysics.

11.4 The Problem of Theory Change in Science

In the philosophy of science, the 'problem of theory change' is the problem of maintaining that the deep-theoretical portions of our current best theories have epistemic as well as instrumental value, given that theories developed to describe any given domain have tended to undergo successive and seemingly violent ruptures. Since it is a debate over the epistemic standing of theories, it is a debate over *realism*. For typically, to assert that one is a 'realist' is, within the philosophy of science, to assert that one believes either that our current best theories are either *approximately true* of both observable and deep-theoretic domains, or, at the very least, that we are *making progress* towards creating theories that are.[31] In neither case is it necessary to hold that our current best theories are right in their every last detail, or that they will never be superseded; but it *is* to hold that, at the very least, non-trivial relations of approximation will exist between our current theories and their successors. For even if our current best theories are too far from the truth to be reasonably regarded as 'approximately true', it seems that only if our theories can be regarded as getting *more* approximately true that they may be arranged into an arc of *progress*. And while it is not entirely obvious what precisely the significance is of mere progress towards truth in the context of a debate over what we can with justification believe, it can hardly be denied that progress would represent a good, something of epistemic value.[32]

This picture in which successive theories represent continual improvements or refinements of one another at both the empirical and theoretical levels is variously

[31] See Worrall (1989, 106). [32] Cf. Worrall (1989).

referred to as 'accumulationist' or 'retentionist' realism, or, for Laudan, 'convergent' realism. In essence, this view is that

> the *historical progress* of the mature sciences is largely a matter of successively *more accurate approximations* to the truth about both observable and unobservable phenomena. Later theories typically build upon the (observational and theoretical) knowledge embodied in previous theories.[33]

As is well known, this optimistic vision of the development of science was rudely interrupted in the early 1980s with Laudan's 'pessimistic induction' (though of course the roots were there in Kuhn and others).[34] For what Laudan there claimed was that the 'approximate truth' of a theory, although a fiendishly difficult notion to pin down, must have as a necessary condition that the central theoretical terms of a theory refer. But looking back at the theories of the ether, the caloric, the phlogiston, Newton's action at a distance forces of gravity, the theory of classical particles, and so on, it seems that from the perspective of current theories none of these theories referred at all. Hence it seems we cannot say that past scientific theories are in any sense 'approximately true' relative to their successors, and there is no way that the history of science can be viewed as a sequence of ever better approximations to the truth. As such, rather than something that can, at deep-theoretical levels, be meaningfully regarded as a sequence of incremental improvements, we are left with a picture of mere displacement and change.

In the face of this challenge, it seems the only option for the convergent realist is to try to maintain that there is in fact some relevant but as yet unnoticed respect in which successive theories may indeed be held to approximate each other. While there have been several attempts to maintain this, one influential response came from structuralists.[35] According to them, rather than focusing on what theories ostensibly refer to, we should focus on the structural relationships in which what they putatively refer to is embedded. Focus, among other things, on the *equations* of the theory and you *will* find something preserved as an approximation through episodes of theory change. For what is observed is that the central equations of successor theories may be deformed into *close approximations* of those of previous theories as some relevant parameter is varied (such as velocity, energy, field strength, etc.). And clearly, part of what makes this possible is the fact that the relation of 'close approximation' has a well-defined meaning in this context, given the inherently mathematical nature of that taken to stand in it (the notion of 'approximation' being a quantitative measure of similarity after all). And if

[33] Boyd (1983, 45). [34] Laudan (1981).

[35] Note that securing structural continuity is arguably a necessary component of *any* realist response to Laudan's challenge—theories canonically being identified by their central equations after all.

we are willing to say that equations describe relationships that are out there in the world, then there is, *contra* Laudan, a sense in which the theories may be said to be 'approximately true'—precisely because, from the perspective of the new theory, they get those relationships approximately right. And indeed, in the cases of the most iconic examples of radical theory change—such as from classical to quantum physics, Galilean to special relativistic kinematics, and Newtonian to Einsteinian gravity—strong relations of approximate structural continuity can indeed be maintained.[36]

On the basis of these sorts of case studies, then, realists sharing the structuralist view on approximate truth will hold that convergent realism can after all be maintained, the vicissitudes of the historical record notwithstanding. As such, it *is* the case that current theories may be regarded as approximately true, or at least as making progress towards that. Now, while this claim can of course be contested on both historical and conceptual grounds, let us for present purposes assume it. For if that is the case, then the value of present-day physics, construed as an attempt to describe the fundamental, is thereby secured, and this *even though we do not at present have a fundamental theory.* But if that is so, it might seem there is hope that a present-day metaphysics 'informed' by these theories could be thought of as making progress also. For the transitions canonically cited in defence of convergent realism are the very same transitions under which our metaphysical claims seem to have had to be jettisoned. Since the relevant scientific claims may be argued to be progress through such changes, might not the corresponding metaphysical claims be said to do so too?

It is my impression that it is precisely this conception of the relation between scientific and metaphysical progress that informs the practice of many, if not most, metaphysicians of science. My reasons for thinking this are admittedly largely anecdotal—although one could of course also regard it as implicit in the very practice of engaging in the metaphysics of science at a point in time in which we fully expect our science to undergo significant future change. However, occasionally one can find explicit statements in the literature that this is indeed the vision that naturalistic metaphysicians have in the background. Here, for example, is one by Ladyman and Ross:

> One consequence of naturalism that cannot be avoided is that if our current scientific image of the world changes much, as we suppose it will, then it will then turn out that the best current metaphysics is substantially wrong. Neither we nor anyone else can do better than articulate the best metaphysical picture the current evidence suggests in attempting to sketch the image of the world

[36] See Saunders (1993) for a comprehensive discussion.

> that science presents to us now.... Note that to the extent that metaphysics is closely motivated by science, we should expect to make progress in metaphysics *iff* we can expect to make progress in science. In Chapter 2 [in which they defend epistemic structural realism as a response to the problem of theory change] we indicate at length why we hold fallibilism about science to be compatible with optimism about epistemic progress in science. *This argument carries directly over to scientifically motivated metaphysics.*[37]

This optimistic picture is one in which the metaphysics of science may be expected to march in lockstep with the sciences themselves: according to it, as the latter makes progress we have every reason to think that the former will make progress too. And it is hard to think of what the underlying justification for this picture could be other than that since naturalistic metaphysics is so closely allied to science, success in the latter should 'trickle down' to the former. But however appealing this vision may be to naturalists, and however *prima facie* plausible claims of this sort may sound, it is my contention that this optimisic vision of metaphysics is not one that we can ultimately sustain. For so long as we conceive of progress between metaphysical theories as we do in the case of scientific theories—namely, in terms of increasingly good approximations to the truth—I contend that metaphysical theories cannot be said to instantiate it. This is because the very nature of the claims that metaphysics canonically makes, framed in terms of metaphysical concepts as canonically understood, seems to resist any notion of 'approximation' being predicated of them. And clearly, if we conceive of progress in metaphysics in a way distinct from how we conceive it in science, it becomes correspondingly unclear how progress in the latter could be expected to percolate down to the former. This gives rise to what I will call the *problem of progress in metaphysics*, which I will try to outline now.

11.5 The Problem of Progress in Metaphysics

Central to my sceptical argument is the claim that metaphysical theories, at least as canonically presented and understood, cannot be regarded as being 'approximately true'. As such, successive metaphysical theories cannot be conceived of as 'more' approximately true than their predecessors, and this even if the physical descriptions those metaphysical theories interpret *can* be regarded in this way. To see this, let us return to the conception of theories of metaphysics presented as canonical in Section 11.2 above—namely, the conception of metaphysics as

[37] Ladyman and Ross (2007, 35) (italics added). Expression of a similar thought may be found in Baker (2013, 265).

concerned with questions of *primitive ideology* and *primitive ontology*. Thus theories of metaphysics, as canonically conceived, we will take to be assertions of the form 'Φ is fundamental' or 'Φ is non-fundamental'—where Φ describes either a piece of ontology or a piece of ideology—together with the *justifications* that are offered for these claims.[38] Theories of metaphysics, then, we will take to be claims of the form

(i) Ψ is (non-)fundamental, because χ

where the Ψ and χ may be taken to involve predicates that we regard as distinctively metaphysical. Thus the thesis of ontic structuralism is the claim that it is structure that constitutes the fundamental category, because the fundamental properties are all extrinsic and the cardinality associated with putatively fundamental 'objects' is not an objective feature of the world. The thesis of Humeanism is the claim that modality is not fundamental, since the fundamental properties are all categorical and intrinsic. And the thesis of presentism involves a commitment to the fundamentality of tensed ideology, since it denies the existence of past and future entities that could analyse the truth conditions of such claims in tenseless terms. As we have seen, the truth-values associated with all of these theories—both the justifications and the claim to fundamentality—have been argued to change as science progresses, and what we are now asking is whether the notion of 'approximate truth' may be appealed to in support of the idea that metaphysics makes progress over and above mere change. First on the agenda, then, is stating what such an assertion would even mean as applied to claims of metaphysics. Now, plausibly, what it means for a claim describing some state of affairs to be 'approximately true' is for a claim describing some approximation of that state of affairs to be true *simpliciter*. And it is hard to see what could be meant by "'Φ is (non-)fundamental' is approximately true" other than that Φ itself approximates that which is (non-)fundamental.[39] But as I will argue now, it is hard to see how claims of the form (i) can be regarded as meaningful *qua* approximations, at least as the metaphysical predicates that feature in them are canonically understood.

To see this, consider first the logic of approximation in more familar, scientific contexts. Consider, for example, the claim that 'x is approximately n metres long'. On the assumption that by x being 'approximately' some way we understand that it is *not exactly* that way—as would certainly be appropriate in the theory-change context at least—then the claim 'x is approximately n metres long' entails that x is, strictly speaking, *not* n metres long, and so has some length in metres $n' \neq n$.

[38] We are after all concerned with epistemic progress, and hence with claims to knowledge. And being right about what is fundamental for all the wrong reasons would not constitute knowledge of it.

[39] I cannot argue for this here without getting into the weeds of the concept of fundamentality. However, I take it that it has *prima facie* plausibility at least.

However, it is *also* to say that x may be *treated* as n metres long *from the point of view of our purposes*. For there is no such thing as being 'approximately' some way *simpliciter*: clearly, the same difference between n and n' could be inconsequential in some contexts and ruinous in others.[40] Thus, when we say that something is 'approximately' some way, we are asserting that it is *not in fact* that way, but also that it can be *treated* that way *for our purposes*. (Certainly that would be the case here, where we positively expect to not be exactly right.) But if such predications are to be non-trivial, it surely must be the case that not *all* lengths are ways that x could be said to be 'approximately'.[41] That is, there must be some other length property n'' such that x is *not* n'' metres long, either strictly speaking or just approximately. Thus for talk of approximation, such as 'x is approximately N' for an arbitrary property N, to be meaningful there must be at least three properties in play: (i) one property N that x fails to instantiate strictly speaking but that x may be regarded as having from the point of view of our purposes; (ii) one (perhaps unknown) 'intermediate' property N' that x instantiates strictly speaking and that is sufficiently 'close' to N from the point of view of those same purposes; and (iii) one 'further away' property N'' that x cannot be regarded as instantiating, either strictly speaking or from a practical point of view. Let this requirement on the meaningfulness of predications of approximateness be the 'intermediacy' condition.

If I am right that satisfaction of the intermediacy condition is necessary for making meaningful predications of approximateness, then it is easy to see why physics is routinely able to do so. For physics typically describes systems in quantitative terms—such as in terms of what positions they have, what probabilities they have for being there, and what the determinate values of their determinable kind properties are (e.g. their specific charges and masses). It therefore describes them in terms of a *spectrum of magnitudes*, each one of which may in a clear sense be said to lie *closer to* and *further away* from some chosen other, as is required by (ii). And given that these spectra tend to feature a *great many* values—indeed, typically infinitely many—there will generally be a wealth of properties that our system will *fail* to have, either approximately or strictly speaking, in accord with (iii). However, it is not progress in physics that concerns us at this point, but rather progress in the metaphysics of physics. Thus, we must ask, are the conditions for meaningful ascriptions of approximateness likewise fulfilled in that context?

But here the answer seems to me to be *no*. For metaphysics tends to work by taking the quantitative properties that are handed to us by physics, and funnelling them into second-order, *sui generis* categories that are generally *non-quantitative*, and moreover highly *dichotomous* in character. For rather than predicating

[40] See Duhem (1991, 172).

[41] That content is inversely proportional to what is permitted was of course stressed by Popper (1962).

quantitative properties of systems drawn from a wide (indeed, typically infinitely wide) spectrum of possibilities concerning their properties—such as what their charges are, or how numerous they are, or what the probabilities associated with either of those are—metaphysicians tend to ask whether properties of that sort are *intrinsic*, or whether they are *categorical*, or whether they are *objective*. But these *sui generis* categories, I maintain, tend to be defined in *mutually exclusive and jointly complete pairs*—pairs which, rather than forming a rich spectrum of possible properties, merely bisect the property space. But it seems to me that dichotomous 'either/or' structure that results—one in some sense so crude in comparison with that of physics—makes the notion of 'approximation' in principle inapplicable to the descriptions that we form in terms of these categories.

To see this, let us start first with the definition of 'intrinsic'—a second-order property that, as we have seen, performs crucial work in the defence of both Humean and structuralist metaphysics. For Humeanism requires *all* fundamental properties to be intrinsic, as part of its doctrine that there are fundamentally no necessary connections between distinct existences. Structuralists, by contrast, require that *no* fundamental properties are intrinsic—or, to put the same idea differently, that 'not just some, but all properties of the particles described by quantum theory are extrinsic'.[42] Now, note first that the very way these metaphysical theses are stated already suggests they have a sort of 'all-or-nothing' character that would seem to make them inhospitable to approximation: for the occurence of *just one* fundamental property that respectively is not or is intrinsic will be enough to scupper them. But when it comes to the definition of 'intrinsic' itself we see an analogous, 'all-or-nothing' phenomenon. For while there is no universally accepted formal definition of this notion, the basic concept that formal definitions are intended to capture is widely agreed upon. For a standard, if intuitive, characterization of 'intrinsic property' is a property that "an object has solely in virtue of how it is, independently of its surroundings"; it is a property that may be possessed by an object "regardless of what may be going on outside of itself".[43] In other words, intrinsic properties are those that "possess the second-order feature of stability-under-variation-in-the-outside-world".[44] Now, notice the wholly unqualified nature of these statements: as stated, should *any* variation in the outside world be ruled out by possession of the property, then it will follow that the property is not intrinsic. However, the standard characterization of *extrinsic* properties is simply "those that are not intrinsic".[45] Further, we do not standardly recognize properties *other* than these logical contraries in our metaphysics of instantiation co-dependence.[46] As such, it seems that the standard, second-order predications that we make of physical properties with regard to the dependences

[42] Chakravartty (2012, 204). [43] McKitrick (2003, 265); Cameron (2009, 158).
[44] Yablo (1999, 41). [45] McKitrick (2003); Cameron (2009). [46] See Figdor (2014).

upon their co-instantiation are in terms of *pairs of categories* that are *mutually exclusive and jointly complete*. But this circumstance would seem to prohibit any meaningful notion of 'approximate intrinsicality'. For it leaves no 'middle ground' between intrinsic and extrinsic to allow for the satisfaction of the intermediacy condition: any measure of dependence whatsoever is sufficient, on the usual definitions, to render an intrinsic property its opposite. The only conclusion to draw is that—again, on the standard renderings of the properties we standardly recognize—there is no scope to speak of metaphysical claims regarding property intrinsicality as being 'approximately true'.

A similar logic applies to the notion of a 'categorical property', centrally involved in debates over Humeanism. Making things in one sense more difficult here is the fact that precise, non-metaphorical characterizations of categorical properties are not easy to come by.[47] At the root of this problem is the fact that—since it is part of the point of categorical properties to be maximally 'thin' metaphysically—it is easier to say what such properties are *not* rather than what they are. Hence as Bird puts it, "what we mean by 'categorical' must be understood in negative terms".[48] Thus, categorical properties are usually articulated in terms of their contrast with *essentially dispositional* properties, which are taken to have a much richer (and thus more readily articulable) structure. Witness, for example, that Mumford (as do many others) uses the terms 'categorical' and 'non-dispositional' interchangeably; while for Bird, "to say that a property is categorical is to deny that it is necessarily dispositional".[49] Now, this richer structure involved in essentially dispositional properties is modal structure: for it is taken to be the mark of such properties that they *entail non-trivial counterfactual conditionals*. Thus "traditionally it is suggested that ascriptions of dispositions do, while ascriptions of categorical properties do not, entail certain counterfactual conditionals".[50] Although stating precisely what this means is a matter of ongoing debate, it is universally agreed that such properties, *qua* fundamental posits, are unacceptable to Humeans—since Humeans require all modal ideology, such as that involved in stating counterfactuals, to be analysed away.

Putting all this together, then, the following emerges. First, this is another instance in which the categorizations central to a metaphysical dispute are defined in mutually exclusive and jointly complete pairs: those of which modal ideology essentially features in its definition, and those of which it does not. Thus our standard categorizations again fail to make space for a metaphysical 'middle ground' that would allow us to talk of approximation meaningfully. But further, insofar as metaphysicians categorize properties in terms of the ideology required in a statement of their nature, then the notion of these properties being 'approximately'

[47] Mumford (1998, 75). [48] Bird (2007, 66).
[49] See e.g. Mumford (2006, 477); Bird (2007, 66–67). [50] See Choi and Fara (2018).

some way or other would seem to make no sense even in principle. For since pieces of ideology, whether predicates or operators, are clearly *either present in a statement or not*, it seems there is no space for anything intermediate between the presence or absence of ideology in a definition. (One clearly cannot have, say, 98 per cent of a predicate present in a proposition.) Hence it seems especially clear here that claims according to which some piece of ideology is fundamental or not cannot be said to be 'approximately true'. And though this point has been made in the context of Humeanism, we can anticipate the very same point being made in the cases of other disputes over fundamental ideology—such as that involved in the debates over presentism or the need for primitive identity.

Finally, let us consider the case of a property found in physics being 'objective'. This is something that we encountered in the discussion of the Unruh effect, used to undermine the objective reality of quantum particles in general curved spacetimes—the reason being, we saw, that different observers will disagree about how many such particles there are in any given region of that spacetime. Now, while there is much here that *can* be quantified—in particular, the extent of disagreement between two given observers, travelling with such-and-such relative accelerations—note that what was cited above in the move from disagreement to lack of objectivity was not the extent of disagreement, but rather *the very fact of disagreement itself*.[51] At least to the protagonists in this debate, then, it seems that *any degree* of disagreement is sufficient to declare that particle number is not an objective feature of the world, and hence, on the standard philosophical characterization, not a feature that may be regarded as 'real'. This suggests that our concept of objectivity, at least as routinely deployed in the metaphysics of physics, again has this all-or-nothing character that would again make qualifications of 'approximate' inappropriate. Indeed, discussing a passage of Nozick's *Invariance* concerning laws, Earman states outright that "I do not see that it makes much sense to speak of gradations of objectivity".[52] If that is the case, then the notion of 'approximate objectivity' would not seem to make much sense either.

It seems, then, that the second-order metaphysical predications that we standardly make of physical properties, and at least as we standardly understand them, tend to have a dichotomous character that suggests they do not meaningfully admit of notions of 'approximation'—and this even though the first-order physical properties themselves generally do so readily. In consequence, it is hard to see how the claims we make framed in terms of these predicates could be said to be 'approximately true', hence not even as making successively better approximations to the truth. But this returns us to the problem of what the value of engaging in

[51] While an inertial observer will see no particles, one accelerating at a relative to the first will 'see' a bath of particles with Planck radiation spectrum $kT = \frac{\hbar a}{2\pi c}$.

[52] Earman (2004, 1233).

metaphysics in the present consists in, given the historical facts of metaphysical theory change together with the fact we expect such changes in the future. For it leaves us bereft of any obvious way of stating how it is that metaphysics makes progress. After all, if metaphysical claims can only be said to be true or false *simpliciter*, and we furthermore expect our judgement of that semantic value to undergo change, then there is no more sense to the idea that metaphysics is 'converging' upon truth any more than that the series $1 - 1 + 1 - 1 + 1 \ldots$ is converging upon a number. Certainly—and *contra* Ladyman and Ross—there is no remotely obvious inference from the fact that physics makes progress to the claim that metaphysics does too. For even when the metaphysics in question is 'informed' by the relevant science, the two notions of progress involved must in each case be quite distinct.

This, then, is the axiological challenge that I put to naturalistic metaphysics—more specifically, to naturalistic metaphysicians making claims about the fundamental while still bereft of a truly fundamental theory. Needless to say, however, the argument has its fair share of weaknesses, and of course there are rejoinders that could be made to it. The most obvious is to simply deny that progress in metaphysics must be conceived of in close analogy to how we conceive of science. Indeed, one might think that is just a crude scientism that would assume that that had to be the case. But if that is so, then we are owed an alternative account of how progress in metaphysics ought in fact to be conceived of, and then an argument to the effect that the metaphysics of physics exhibits it. While that account would be welcome, for all sorts of reasons, I have yet to see it made. Another route, however, is to deny that canonical metaphysical categories are necessarily such as to be unamenable to notions of approximation. For much of the above argument turned on mere observation of how such terms are understood, without providing—at least not in all cases—a principled reason why they must be understood in that way. Thus while it may be the case that our notions of 'intrinsic' or 'objective' *as currently understood* make them unreceptive to notions of approximation, there is no reason in principle why we could not understand them differently. Perhaps, for example, a notion of 'approximate intrinsicality' may in fact be meaningfully defined.[53]

This is surely an interesting question. Note, however, that since the notion of approximation is always tied to our purposes, it probably cannot be answered without asking ourselves what the purpose served by metaphysics is. (What, for example, do we hope to achieve by arguing that structuralism, or Humeanism, correctly reflects the fundamental nature of the world? What do we hope to gain by establishing this?) While it is unclear to me where one even begins to

[53] Though I will not discuss the grounds or content of these positions here, it should be noted that Cassirer (1923, chapter 6), in addition to Nozick (2001), talks of 'degrees of objectivity'.

start answering it, perhaps it is a question which we, as naturalists, ought to be thinking about independently. Indeed, it seems unlikely that any alternative account of progress in metaphysics, needed to answer the axiological challenge, can be developed without engaging with it at some point at least.

11.6 Conclusion

It seems to me that the naturalistic metaphysician faces the following challenge. As metaphysicians, we seem to dearly want to make claims about the fundamental, but at present we lack the theory required to inform those claims. The best we can do at the moment is say what is fundamental on the basis of our current theories, and those we expect to change. But we expect our metaphysics to change alongside those theories: what seems true today will in all likelihood appear false in the future. And nor, I have argued, is there any clear sense in which the claims we make today can be said to be approximately true, or even as 'more' approximately true than before. It is therefore unclear how we may be said to be making progress in metaphysics if understood in terms analogous to those in which we understand progress in science. As such, the value of engaging in metaphysical speculation at present emerges as deeply unclear. Given that naturalists have spent much of their energy denigrating the metaphysics of others, it strikes me that we need to find some justification for thinking that our project, at least as pursued in the present, enjoys a more elevated status.

References

Baker, D. J. (2013). Identity, superselection theory, and the statistical properties of quantum fields. *Philosophy of Science 80*(2), 262–285.

Bhogal, H. and Z. Perry (2017). What the Humean should say about entanglement. *Noûs 51*(1), 74–94.

Bird, A. (2007). *Nature's Metaphysics: Laws and Properties*. Oxford: Oxford University Press.

Bitbol, M. (2007). Schrödinger against particles and quantum jumps. In J. Evans and A. S. Thorndike (Eds.), *Quantum Mechanics at the Crossroads*, pp. 81–106. New York: Springer.

Boyd, R. N. (1983). On the current status of the issue of scientific realism. *Erkenntnis 19* (1–3), 45–90.

Brading, K. and E. Castellani (Eds.) (2003). *Symmetries in Physics: Philosophical Reflections*. Cambridge: Cambridge University Press.

Cameron, R. (2009). Intrinsic and extrinsic properties. In R. Le Poidevin, P. Simons, A. McGonigal, and R. Cameron (Eds.), *Routledge Companion to Metaphysics*. London: Routledge.

Cassirer, E. (1923). *Substance and Function and Einstein's Theory of Relativity*. Chicago, IL: Open Court.

Chakravartty, A. (2012). Ontological priority: The conceptual basis of non-eliminative, ontic structural realism. In E. Landry and D. Rickles (Eds.), *Structural Realism: Structure, Object, and Causality*, pp. 187–206. New York: Springer.

Choi, S. and M. Fara (2018). Dispositions. In E. N. Zalta (Ed.), *The Stanford Encyclopedia of Philosophy* (Fall 2018 ed.). Metaphysics Research Lab, Stanford University. https://plato.stanford.edu/archives/fall2018/entries/dispositions/.

Crispino, L. C., A. Higuchi, and G. E. Matsas (2008). The Unruh effect and its applications. *Reviews of Modern Physics 80*(3), 787–838.

Dalla Chiara, M. L. and G. T. Di Francia (1993). Individuals, kinds and names in physics. In G. Corsi, M. L. Dalla Chiara, and G. C. Ghirardi (Eds.), *Bridging the Gap: Philosophy, Mathematics, and Physics*, pp. 261–283. New York: Springer.

Duhem, P. (1991). *The Aim and Structure of Physical Theory*. Princeton, NJ: Princeton University Press.

Earman, J. (2004). Laws, symmetry, and symmetry breaking: Invariance, conservation principles, and objectivity. *Philosophy of Science 71*(5), 1227–1241.

Figdor, C. (2014). What's the use of an intrinsic property? In R. Francescotti (Ed.), *Companion to Intrinsic Properties*, pp. 139–156. Berlin: de Gruyter.

Finkelstein, D. R. (2006). Whither quantum theory? In K. Williams (Ed.), *Two Cultures: Essays in Honour of David Speiser*, pp. 25–38. Basel: Birkhäuser.

French, S. (2012). Unitary inequivalence as a problem for structural realism. *Studies in History and Philosophy of Modern Physics 43*(2), 121–136.

French, S. and D. Krause (2006). *Identity in Physics: A Historical, Philosophical, and Formal Analysis*. Oxford: Oxford University Press.

Friedman, M. (1983). *Foundations of Space-Time Theories Relativistic Physics and Philosophy of Science*. Princeton, NJ: Princeton University Press.

Hawking, S. (1998). *A Brief History of Time*. New York: Bantam Books.

Huggett, N. (1997). Identity, quantum mechanics and common sense. *The Monist 80*(1), 118–130.

Ismael, J. and J. Schaffer (2016). Quantum holism: Nonseparability as common ground. *Synthese*. https://doi.org/10.1007/s11229-016-1201-2.

Ladyman, J. and D. Ross (2007). *Every Thing Must Go: Metaphysics Naturalized*. Oxford: Oxford University Press.

Langton, R. and D. Lewis (1998). Defining 'intrinsic'. *Philosophy and Phenomenological Research 58*(2), 333–345.

Laudan, L. (1981). A confutation of convergent realism. *Philosophy of Science 48*(1), 19–49.

Lewis, D. (1998). *Papers in Metaphysics and Epistemology*, p. 222. Cambridge: Cambridge University Press. https://www.cambridge.org/core/books/papers-in-metaphysics-and-epistemology/951D5B7E172A80A173E0850D6CF8070E

Lewis, P. J. (2016). *Quantum Ontology: A Guide to the Metaphysics of Quantum Mechanics*. New York: Oxford University Press.

Loewer, B. (1996). Humean supervenience. *Philosophical Topics 24*(1), 101–127.

Lowe, E. J. (2001). *The Possibility of Metaphysics: Substance, Identity, and Time*. Oxford: Clarendon Press.

Maxwell, G. (1970). Structural realism and the meaning of theoretical terms. In M. Radner and S. Winokur (Eds.), *Analyses of Theories and Methods of Physics and Psychology*, Volume 4, pp. 181–192. Minneapolis, MN: University of Minnesota Press.

McKenzie, K. (2016). Looking forward, not back: Supporting structuralism in the present. *Studies in History and Philosophy of Science Part A 59*, 87–95.

McKenzie, K. (2020). Structuralism in the idiom of determination. *British Journal for the Philosophy of Science 71*(2), 497–522.

McKitrick, J. (2003). A case for extrinsic dispositions. *Australasian Journal of Philosophy 81*(2), 155–174.

Monton, B. (2011). Prolegomena to any future physics-based metaphysics. In *Oxford Studies in Philosophy of Religion*, Volume 3, pp. 1–24. New York: Oxford University Press.

Mumford, S. (1998). *Dispositions*. Oxford: Oxford University Press.

Mumford, S. (2006). The ungrounded argument. *Synthese 149*(3), 471–489.

Myrvold, W. C. (2015). What is a wavefunction? *Synthese 192*(10), 3247–3274.

Nozick, R. (2001). *Invariances: The Structure of the Objective World*. Cambridge, MA: Harvard University Press.

Putnam, H. (1967). Time and physical geometry. *Journal of Philosophy 64*(8), 240–247.

Quine, W. V. (1951). Ontology and ideology. *Philosophical Studies 2*(1), 11–15.

Saunders, S. (1993). To what physics corresponds. In S. French and H. Kamminga (Eds.), *Correspondence, Invariance and Heuristics: Essays in Honour of Heinz Post*, pp. 295–325. New York: Springer.

Saunders, S. (2003). Physics and Leibniz's principles. In K. Brading and E. Castellani (Eds.), *Symmetries in Physics: Philosophical Reflections*, pp. 289–307. Cambridge: Cambridge University Press.

Sider, T. (2001). *Four-Dimensionalism: An Ontology of Persistence and Time*. Oxford: Oxford University Press.

Sider, T. (2013). *Writing the Book of the World*. Oxford: Oxford University Press.

Wald, R. M. (1994). *Quantum Field Theory in Curved Spacetime and Black Hole Thermodynamics*. Chicago, IL: University of Chicago Press.

Weinberg, S. (1994). *Dreams of a Final Theory*. New York: Vintage.

Williams, P. (2019). Scientific realism made effective. *British Journal for the Philosophy of Science 70*(1), 209–237.

Wilson, J. M. (2015). Hume's dictum and metaphysical modality: Lewis's combinatorialism. In B. Loewer and J. Schaffer (Eds.), *Blackwell Companion to David Lewis*, pp. 138–158. Oxford: Blackwell.

Worrall, J. (1989). Structural realism: The best of both worlds? *dialectica 43*(1–2), 99–124.

Wüthrich, C. (2010). No presentism in quantum gravity. In V. Petkov (Ed.), *Space, Time, and Spacetime: Physical and Philosophical Implications of Minkowski's Unification of Space and Time*, pp. 257–278. Berlin: Springer.

Yablo, S. (1999). Intrinsicness. *Philosophical Topics 26*(1/2), 479–505.

12

Problems with the Cosmological Constant Problem

Adam Koberinski

Abstract

The cosmological constant problem is widely viewed as an important barrier and hint to merging quantum field theory and general relativity. It is a barrier insofar as it remains unsolved, and a solution may hint at a fuller theory of quantum gravity. I critically examine the arguments used to pose the cosmological constant problem, and find many of the steps poorly justified. In particular, there is little reason to accept an absolute zero-point energy scale in quantum field theory, and standard calculations are badly divergent. It is also unclear exactly how a semiclassical treatment of gravity would include a vacuum energy contribution to the total stress-energy. Large classes of solution strategies are also found to be conceptually wanting. I conclude that one should not accept the cosmological constant problem as a problem that must be solved by a future theory of quantum gravity.

12.1 Introduction: What's the Problem?

There is a near consensus among theoretical physicists working on a quantum theory of gravity—particularly those trained in particle physics—that the cosmological constant poses a major theoretical problem.[1] That's where the consensus ends; exactly *what* the problem is and *how* it affects our understanding of physics is unclear. One thing is clear: the problem involves an apparent conflict between quantum field theory and the observed value of the cosmological constant needed for our best models of the universe. In his seminal review, Weinberg (1989) introduces the cosmological constant problem (CCP) as one of the few crises available to serve as fodder for theoretical physicists:

[1] There is not *complete* consensus, however. See Bianchi and Rovelli (2010) for one notable criticism of the idea that the cosmological constant poses any problem at all.

Adam Koberinski, *Problems with the Cosmological Constant Problem* In: *Philosophy Beyond Spacetime: Implications from Quantum Gravity*. Edited by: Christian Wüthrich, Baptiste Le Bihan, and Nick Huggett, Oxford University Press.
 DOI: 10.1093/oso/9780198844143.003.0012

> Perhaps it is for want of other crises to worry about that interest is increasingly centered on one veritable crisis: theoretical expectations for the cosmological constant exceed observational limits by some 120 orders of magnitude. (1)

Others share the worry, but drastically decrease the magnitude of the discrepancy. Burgess (2015) claims that the CCP stems from attempting to answer a simple question: "What is the vacuum's energy, and how does it gravitate?"

> [I]t is a problem because observations indicate a vacuum energy that is more that 50 orders of magnitude smaller than we think it should be, given the particles and interactions that we find in experiments. (1)

Others take a different approach to stating the problem. Rather than focus on the magnitude of the discrepancy between prediction and observation, one may wonder why the observed cosmological constant is so small, why it isn't exactly zero, or why the vacuum energy density is on the same order of magnitude as the current matter energy density of the universe (see Nobbenhuis 2006, section 1).[2] Still others think that all of these are ill stated, and don't get at the heart of the CCP:

> In some respects, the biggest problem with the cosmological constant problem is that it is rarely stated properly. Why is the cosmological constant so small? Why is it not Planckian? Why is it not zero? All of these questions belittle the cosmological constant problem, and if we are to have any hope of solving it, we had better be clear about what the problem really is…[T]he real issue with the cosmological constant is not so much one of fine tuning, but of radiative instability, and the need to *repeatedly* fine tune whenever higher loop corrections are included. (Padilla 2015, 2)

Radiative instability is uniquely a problem for perturbative formulations of the CCP; order-by-order fine-tuning only occurs in a perturbative treatment. A small minority don't think that there even is such a thing as the CCP. Bianchi and Rovelli (2010) are highly critical of stating the CCP as a crisis for current physics:

> There is no 'great mystery', in an accelerated expansion of the universe. This is a phenomenon which is clearly predicted and simply described by well-understood

[2] A brief comment on the nature of the citations given here. I focus on review articles and lecture notes from physicists. Lecture notes are particularly illuminating, since they are used to convince the 'uninitiated' graduate students that the problem really is a problem. Given the near-universal acceptance of the CCP among working physicists, these arguments are often unstated in published articles. The review articles are an exception—they often feature critical analyses of proposed solutions, as well as somewhat detailed discussion of the nature of the CCP itself.

> current physical theory. It is well understood in the context of general relativity, which naturally includes a cosmological constant...the common theoretical objections against this interpretation of the acceleration are either weak, or ill-founded. (1)

So, with the disagreement amongst physicists about what the problem is, or whether there is a problem at all, it is no surprise that the CCP remains a confusing ground for philosophers. Saunders (2002) has examined the philosophical issues concerning the reality of the quantum vacuum, in analogy with the electromagnetic aether and the Dirac negative energy sea. In the latter cases, physics managed to do away with these concepts while maintaining phenomena they were originally thought to explain. The technical issues with the vacuum in quantum field theory are a bit tougher, however. Rugh and Zinkernagel (2002) provide the first detailed philosophical analysis of the CCP as a whole, and critically highlight some of the conceptual barriers I discuss below. Where they remain largely agnostic regarding the resolution of these barriers, I argue that they provide good reason to reject the CCP.

In section 12.1.1, I will provide the relevant background to the CCP, and attempt to disambiguate the seemingly distinct issues that get lumped together under its umbrella. I take all of these to contribute to *the* CCP, but each is worth mentioning as a separate piece. Section 12.2 outlines the background assumptions that one must accept in order to accept the CCP as a real problem. In subsections I detail the steps, and highlight philosophical issues with each. The steps build upon one another; if one rejects any step—as I will argue one should—then the CCP doesn't arise at all. Section 12.3 puts these issues aside, and outlines some of the broad strategies for solving some form of the CCP, given that one takes it to be a problem in the first place. Though interesting models have been proposed in response to the CCP, none adequately deal with the problem. Coupling the lack of solution with the shaky motivation, I conclude that one should not take the CCP seriously as an obstacle to future theories of quantum gravity.

12.1.1 Multiple CCPs

From the previous section, we learn that physicists often speak of different—though related—problems using the phrase 'the cosmological constant problem'. In fact, introductions to the problem will often list three distinct problems, commonly referred to as the old CCP, the new CCP, and the coincidence problem. The first two are more appropriately thought of as challenges for future models of quantum gravity, rather than issues with the current Standard Model of particle physics of the ΛCDM model of cosmology. Our best theory of matter—quantum field theory—seems to tell us that the vacuum has an enormous energy density. Our best theory of gravity—general relativity—tells us that this energy density should gravitate, and a semiclassical merger of the two indicates that vacuum energy density should couple to the Einstein field equations in the form of a cosmological constant term. What we get from quantum field theory is an enormous cosmological constant term, well beyond what would be physically allowable based on observation.[3] Requiring that the quantum vacuum be Lorentz invariant heavily constrains the form of the vacuum stress-energy tensor. The most general tensor quantity that is Lorentz invariant is a multiple of the Minkowski metric, so $\langle T^{\mu\nu}\rangle \propto \eta^{\mu\nu}$. Through the standard process of minimal gravitational coupling, we replace the Minkowski metric with a general Einstein metric, and introduce the expectation value of the vacuum energy as the correct constant of proportionality,

$$\langle T_{vac}\rangle^{\mu\nu} = -\langle \rho_{vac}\rangle g^{\mu\nu}, \tag{1}$$

where conservation of stress-energy forces $\langle \rho_{vac}\rangle$ to be constant.[4] Equation 1 looks like a special case of the stress-energy of a perfect fluid, where $p_{vac} = w\rho_{vac}$, indicating that the equation of state for the vacuum is parameterized by $w = -1$. If we insert $\langle T_{vac}\rangle^{\mu\nu}$ into the Einstein field equations, it takes on a form identical to that of the cosmological constant, with $\Lambda_{vac} = 8\pi G\langle \rho_{vac}\rangle$. By taking the vacuum energy seriously, we seem to 'predict' the presence of a cosmological constant in a semiclassical treatment of gravity.

The old CCP assumes that $\Lambda = 0$, and leads to attempts to find mechanisms going beyond quantum field theory and general relativity for eliminating the vacuum energy contribution. However, observations have since indicated that the expansion of the universe is accelerating, meaning that our best cosmological models now include a nonzero Λ. Thus, the new CCP is a bit more challenging: one must find a way of canceling the vacuum contributions, while also including a small, positive effective Λ. Cancellations in light of the new CCP need to be finely tuned to at least fifty orders of magnitude in order to leave a residual $\Lambda \neq 0$, which physicists find unnatural.[5]

Both the old and new CCPs are made more severe by the fact of radiative instability. Unlike certain other quantities in quantum field theory—such as the mass and charge of the electron in quantum electrodynamics—the diverging value of the

[3] The discrepancy is so large that the very fact of our existence would preclude a cosmological constant term anywhere near that predicted. Weinberg (1989) famously 'predicted' the now observed value of the cosmological constant by placing anthropic bounds on its value.

[4] While $\langle \rho_{vac}\rangle$ must be constant, we should expect ρ_{vac} to vary heavily. The quantum vacuum fluctuates, and these fluctuations are supposed to provide some empirical evidence for the reality of vacuum energy. A fully quantum treatment would force ρ_{vac} to be far from constant.

[5] It's not obvious why a high degree of cancellation resulting in $\Lambda = 0$ is any more natural than one resulting in $\Lambda \neq 0$, but theorists' intuitions seem to make the new CCP even more pressing.

cosmological constant is not properly renormalizable. Renormalization involves 'redefining' constants in a theory in such a way that divergences cancel out. A renormalizable term is cured of divergences at a finite order within a perturbative expansion, and remains cured at all higher orders. The vacuum energy in quantum field theory is radiatively unstable, which leads to the traditional forms of the CCP. Radiatively unstable terms can be tuned to a given order in a perturbative expansion, but they diverge again when higher-order terms are included. If the vacuum energy were renormalizable, there would be no CCP at all. One would simply redefine the vacuum energy as the physically observed vacuum energy, and this would be no more puzzling than the mass or charge of the electron. One may be unsatisfied with the renormalization procedure, but this is a separate issue from the CCP. Though not explicitly stated in many formulations of the CCP, the radiative instability of the vacuum energy is extremely important for understanding problems with fine-tuning solutions.

Given the radiative instability of the vacuum energy, a finite value is obtained by placing an explicit cutoff in the integrals of the theory. Though this step is often taken in quantum field theory, renormalizable terms end up without a dependence on the exact value of the cutoff used. The vacuum energy density, by contrast, scales at high energies like σ^4, where σ is the cutoff scale. Though the explicit cutoff value is somewhat arbitrary, some value must be chosen to get a quantitative prediction for the vacuum energy density.[6] The fact that the problem persists across any physically reasonable choice of cutoff is used as justification for the robustness of the problem: even though we don't know the correct cutoff scale for Standard Model quantum field theory, the CCP will arise at any value of the cutoff.

The coincidence problem, which is often grouped together with the CCP, is a rather distinct problem, independent of considerations from quantum field theory. In cosmological models the total stress-energy tensor is dominated by various contributions at different time scales. In the early universe, radiation dominates. But because there is a finite amount of radiation, and its energy density in a region of radius r scales as r^{-4}, the radiation dilutes quickly, and the universe moves into a matter-dominated stage. The matter-dominated stage is one in which structures can form, and beings like us can evolve. Matter energy density dilutes as r^{-3} and so this epoch is also temporary. Finally, the energy density from the cosmological constant is (unsurprisingly) constant, and the final epoch is dominated by Λ. This is an era of mostly empty space, rapidly expanding and diluting the matter and radiation present in the universe. Based on observation

[6] The size of the discrepancy in the CCP depends on the exact cutoff used. Weinberg gets a 120-orders-of-magnitude disagreement by taking the cutoff up to the Planck scale, while taking the cutoff to just above energies accessible at the Large Hadron Collider (LHC) gives something like a disagreement of fifty orders of magnitude.

and our best cosmological models, matter contributions (including dark matter) make up just over one quarter of the energy in the universe, while dark energy (in the form of a cosmological constant) makes up the remainder. Thus, it appears we are in an era where the matter and vacuum energy contributions are of the same order of magnitude. If we were to pick a time interval at random from the life of the universe, this would seem an improbable era in which to exist.[7] Even factoring in that we must live in an era with sufficient matter, it seems odd then that we are in a transition era between matter dominated and dark energy dominated. A convincing solution to this coincidence problem would make it less surprising that we find ourselves in this current era, and not some other. This is a mostly unrelated problem, but it does involve the observed magnitude of the cosmological constant, so it often falls under the umbrella of the CCP. As such, I will put the coincidence problem aside for the purposes of this chapter.

12.2 The Steps Required to Accept a CCP

In order to accept the CCP as a problem, one must accept a series of implicit steps. These are rarely mentioned in presentations of the problem, and when they are, it is only in passing.[8] First, one must accept a realism about the vacuum energy in quantum field theory (section 12.2.1). Next, one must trust some 'prediction' of the expectation value of vacuum energy density from quantum field theory (section 12.2.2). This must then be combined with the Einstein field equations in a suitable way (section 12.2.3). Finally, we have to treat the vacuum energy contribution as the dominant contribution to the cosmological constant (section 12.2.4). In the first three of the four steps, there are convincing reasons for rejecting the orthodoxy and therefore rejecting the CCP. Regardless of the heuristic purpose that the CCP serves in constraining future theories of quantum gravity, it is ill motivated from the point of view of quantum field theory, and questionable when quantum field theory is stitched together with general relativity. This section will

[7] There is a rich philosophical literature on issues like this, and whether likelihoods are being appropriately applied within cosmology. White (2000); Norton (2010); and McCoy (2018) are all rather critical of the use of probabilities or likelihoods in these cosmological settings, while Kotzen (2012) and Benétreau-Dupin (2015) defend their standard use, with slight modifications.

[8] For example, Padilla (2015) highlights the Lamb shift and Casimir effect as evidence in favour of accepting step 1 (below). In a footnote, he notes:

> However, as Jaffe (2005) has emphasized, the calculation of both of these effects correspond to Feynman diagrams with external legs of Standard Model fields. A true vacuum bubble has no external legs at all, and from a gravitational perspective the diagrams we are interested in only have external graviton legs. (4)

This footnote acknowledges that there is controversy over accepting the CCP, but is not mentioned again in the rest of the lecture notes.

proceed as follows. I will list a step necessary to arrive at the CCP, present a standard argument for it, and raise some objections to its acceptance. Since the steps build off one another, the objections are independent. That is, if you reject my objections to step 1, then continue on to step 2. If you accept any of the objections, then that suffices to reject the CCP in full.

12.2.1 Is the Vacuum Energy Real?

In quantum field theory on Minkowski spacetime, a particular dynamical model is described by an action. For simplicity, I will work with a free scalar field in an external potential. Nothing regarding a vacuum energy term depends on this choice, except where noted. The action for a free scalar field ϕ in external potential V is given by

$$S = \int \mathrm{d}^4x \left[V(\phi) + \frac{1}{2}\left(\partial\phi\right)^2 \right]. \tag{2}$$

The energy density of the field is given by the Hamiltonian density operator

$$\mathcal{H} = V(\phi) + \frac{1}{2}(\partial_t\phi)^2 + \frac{1}{2}\sum_i(\partial_i\phi)^2,$$

where the index i runs over the spatial coordinates. This Hamiltonian density is minimized at some constant field configuration ϕ_0 (depending on the particular form of the potential V). The leading quantum corrections to the Hamiltonian density come from a quadratic expansion of the excitations about ϕ_0 in terms of creation and annihilation operators,[9]

$$\hat{\phi}(x) := \phi(x) - \phi_0 = \sum_k \left[u_k(x)\hat{a}_k + u_k^*(x)\hat{a}_k^\dagger \right]. \tag{3}$$

Putting equation 3 into the Hamiltonian density and integrating over the spatial coordinates gives the total energy operator

$$\hat{H} = \left[\frac{1}{2}\sum_k \omega_k + V(\phi_0)\int \mathrm{d}^3x \right] + \sum_k \omega_k \hat{a}_k^\dagger \hat{a}_k, \tag{4}$$

[9] Strictly speaking, we must put the field theory in a finite volume first. As we take the volume to infinity, the sums below become integrals over continuous k, and technically the operators become unbounded. These issues are glossed over in most of the physics literature, especially in presentations of the CCP, so I neglect them here.

where the term in square brackets is the zero point, or vacuum energy associated with $\hat{\phi}$, and $\partial_t u_k(x) = -i\omega_k u_k(x)$, $\omega_k \geq 0$. Note that the vacuum energy as given in equation 4 is divergent in two ways. First, one sums over ω_k for all k. This is analogous to summing the ground state energy of an infinite collection of harmonic oscillators, indexed by k. Second, as one takes the volume of space to infinity, the integral over space diverges linearly. For most purposes in quantum field theory, this infinite zero-point energy is simply ignored. We are often interested in expectation values of excited states, and the relevant energy quantities are *differences* between the excited and ground states. The vacuum energy term is simply discarded, and the energy operator is defined as $\hat{H} := \sum_k \omega_k \hat{a}_k^\dagger \hat{a}_k$.[10]

In terms of path integrals and scattering amplitudes in quantum field theory, vacuum energy terms appear as bubble diagrams in the perturbative expansion. In calculating scattering amplitudes, bubble diagrams appear as a product term multiplying all diagrams with external matter lines in the numerator, and so cancel out the analogous bubble diagrams in the denominator. Thus true vacuum energy terms never appear in calculations in quantum field theory.

If we don't ever use it, what grounds do we have for taking the vacuum energy in quantum field theory to be anything more than an inconvenient feature of quantization that we rid ourselves of when calculating? As mentioned above,[8] the Casimir effect and the Lamb shift are often mentioned as effects that force us to take the vacuum energy seriously. I will start with the Casimir effect, as typical presentations appear to depend explicitly on the vacuum state. Casimir (1948) first predicted that closely spaced, uncharged parallel plates would feel a small attractive force, apparently due to fluctuations in the electromagnetic vacuum energy. If one approximates the plates as infinite, impenetrable boundaries, then the spontaneous creation of virtual photons[11] out of the vacuum is constrained between the two plates, but not outside. Since the plates are impenetrable, only photons of appropriate wavelength (i.e., those for which the plate separation is an integer multiple of their wavelength) can be created between the plates. This causes a decrease in energy density in the gap between the plates, and therefore a decrease in pressure. The plates then move toward each other with a small attractive force. The effect was measured by Sparnaay (1957), seemingly verifying the reality of quantum vacuum fluctuations.

[10] In the context of a scalar field, the 'subtraction' of vacuum energy can be justified by demanding that the operators in our theory be normal ordered. Roughly, this means that all operators are redefined such that creation operators are always to the right of annihilation operators. However, normal ordering does not work for more realistic gauge theories.

[11] Virtual particles are short-lived fluctuations that go undetected by normal means. In addition to their rapid annihilation, virtual particles are not constrained to obey mass-energy relations, and are therefore often called 'off mass-shell'.

Jaffe (2005) has called into question this orthodoxy, however. He re-derives the Casimir effect in terms of relativistic forces between charges within the plates and current couplings between them.

> [T]he Casimir effect gives no more (or less) support for the reality of the vacuum energy of fluctuating quantum fields than any other one-loop effect in quantum electrodynamics, like the vacuum polarization contribution to the Lamb shift, for example. The Casimir force can be calculated without reference to vacuum fluctuations, and like all other observable effects in [quantum electrodynamics], it vanishes as the fine structure constant, α, goes to zero. (1)

On Jaffe's construction, the Casimir effect does not depend on vacuum pressure differences, but simply on loop diagram corrections. This is also a more general construction, since different geometric configurations can lead to repulsive forces between conducting plates, and these are not explainable by appeal to the vacuum pressure. It is clear at least that the Casimir effect need not be thought of as a differential vacuum pressure on the parallel plates. Following Jaffe, the Casimir effect should not be treated as any more special than the Lamb shift. Now I will turn to the Lamb shift, and argue that it—along with other one-loop correction effects—should not be taken as evidence for the reality of vacuum energy.

Lamb and Retherford (1947) discovered a small difference in the energy of 2s and 2p orbitals in the hydrogen atom not predicted by the Dirac equation. In quantum electrodynamics, the effect can be seen in a one-loop correction to the interaction between the hydrogen's proton and electron. The one-loop correction is often interpreted as a polarization of the vacuum in response to the interaction between proton and electron. This polarization causes an energy shift in the effective potential, proportional to the square amplitude of the electron's wavefuntion at the center of the atom. Since all electronic orbitals but the s-orbitals vanish at the origin, the shift only affects orbitals with angular momentum $l = 0$, $\Delta E \approx (4m_e)/(3\pi n^3) ln(1/\alpha)\delta_{l,0}$.

Does vacuum polarization force us to take the magnitude of the quantum vacuum energy seriously? Recall that the magnitude of the vacuum energy is represented pictorially by the bubble diagrams in a perturbative expansion. Bubble diagrams are Feynman diagrams with *no external matter lines*, and are therefore (by definition) undetectable. What I have been referring to as vacuum polarization effects are at best evidence for fluctuations in the vacuum energy in response to ordinary matter and its interactions. At worst, one might caution against reading Feynman diagrams too literally; then the vacuum fluctuations as represented in loop diagrams might be interpreted as simply components of nonlinear interaction effects between ordinary matter. Effects like the Lamb shift and Casimir effect give evidence for nothing more than energy fluctuations within the vacuum. The actual magnitude of energy contained in the vacuum has no evidential support. Even within quantum field theory itself, there is no reason to trust the prediction of expected vacuum energy, as I will argue next.

12.2.2 How Much Energy Is in the Vacuum?

Suppose we reject the arguments against the zero-point energies presented in the last section. Given that the vacuum contribution to equation 4 is divergent, how should we determine the magnitude of the vacuum energy density? In order to get a prediction to compare to cosmological observation, we need some way of managing the divergence. Standard renormalization procedures introduce a cutoff scale into the momentum integrals in scattering amplitudes. As long as the divergent quantities (e.g., charge and mass in quantum electrodynamics) scale logarithmically with the cutoff, the divergences can be safely absorbed into other parameters, and the cutoffs can be taken to infinity with the newly redefined physical masses and charges, for example. This procedure fails for the divergent vacuum energy, however.[12] Vacuum energy corresponding to a field of mass m scales with σ^4, where σ is the cutoff scale:

$$\rho_{vac} \propto \int_0^{\sigma} \sqrt{k^2 + m^2}\, k^2 dk \approx \sigma^4. \tag{5}$$

Estimates of the magnitude of vacuum energy density from quantum field theory start by assuming a reasonable cutoff energy scale, up to which we trust the Standard Model. This is typically taken to be the TeV scale, so 'predictions' of the vacuum energy's contribution to the cosmological constant are on the order of at least $(TeV)^4$ It turns out that the exact scale at which the cutoff is introduced is irrelevant—for *any* realistic cutoff scale, $\rho_{vac} \gg \Lambda_{observed}$. The robustness of the problem under changes to the cutoff is supposed to indicate that the problem is somehow independent of the cutoff.

The problem here lies in the fact that any calculation of the vacuum energy cannot be written in a cutoff-independent fashion. Given the more modern view of the Standard Model as an effective field theory, strict renormalizability might not necessarily be required to have a well-defined or physically meaningful vacuum energy. At higher energies, nonrenormalizable terms in an effective field theory are suppressed by some characteristic energy factor, or by the introduction

[12] Though more sophisticated renormalization techniques exist, such as Wilson and Kogut's (1974) lattice renormalization and 't Hooft and Veltman's (1972) dimensional regularization, none succeed in renormalizing the vacuum energy.

of counterterms in the Lagrangian. However, given our ignorance of beyond Standard Model physics, there are no principled grounds for assuming that the divergences of ρ_{vac} will generically be cured in higher-energy theories. Any convincing argument supporting suppression will assume a particular class of high-energy solution, and this is problematic if the CCP is supposed to be a problem for a simple merger of gravity with the Standard Model. The CCP is supposed to be a generic problem, and if a particular class of successor theories is required to even state the problem, one may naturally question its universality, as well as its relevance to current well-accepted theories.

Further, the effective field theory framework assumes a clean separation of physics at differing energy scales (Weinberg 1999). When physics separates cleanly at different scales, high-energy dynamics have little effect on low-energy dynamics, typically in the form of fixing certain 'constants' in the low-energy domain. These can be empirically determined, and then the high-energy theory is screened off from the low-energy domain of interest. This separation of scales works very well for local matter theories, but some question its applicability in cosmological contexts, particularly those involving a global cosmological constant (Afshordi 2015). As I will discuss below, certain classes of proposed solutions also violate separation of scales, and therefore the motivation for the CCP is undercut in a solution.

Historically, divergent terms in a quantum field theory are not trusted until some mechanism for their suppression or renormalization has been demonstrated. I believe that the vacuum energy should be no different; cutoff-dependent values for ρ_{vac} should not be taken as good quantitative predictions, or even estimates for the magnitude of a cosmological constant. This is especially true when alternative approaches to merging quantum field theory and general relativity do not run into such problems.

12.2.3 How Should the Vacuum Energy Enter Einstein's Field Equations?

Suppose, again, that we reject the arguments from the previous two subsections, accept the vacuum energy as real (section 12.2.1), and accept the cutoff-dependent value that is calculated from Standard Model quantum field theories (section 12.2.2). How should we couple the vacuum energy to gravity? As outlined in section 12.1.1, global Lorentz invariance and minimal coupling introduce a vacuum stress-energy contribution of the form shown in equation 1. In Minkowski spacetime, Lorentz invariance of vacuum observables properly constrains $\langle \rho_{vac} \rangle$ to be constant. I will go into a bit more detail regarding the process of minimal coupling here.[13]

[13] Thanks to Michael Schneider for helpful discussion regarding the differences in minimal coupling between general relativity and quantum field theory.

Starting with the Einstein-Hilbert Lagrangian, we add a matter Lagrangian—in particular the vacuum energy density contribution—and introduce an interaction of the form $\alpha R\rho_{vac}$ to couple the two Lagrangians. Minimal coupling is the requirement that the interaction term vanishes, or that $\alpha = 0$. This is the simplest way to introduce the vacuum energy into the Einstein-Hilbert action. Now, gravity only couples to the vacuum energy via the Lorentz-invariant measure $\sqrt{-g}d^4x$:

$$S = \int d^4x\sqrt{-g}\left\{\frac{1}{2\kappa}R + \langle\rho_{vac}\rangle\right\}. \tag{6}$$

Deriving the equations of motion by varying the action leads to a stress-energy tensor of the form $T^{\mu\nu} = -\langle\rho_{vac}\rangle g^{\mu\nu}$. Since minimal coupling works so well in quantum field theory, this appears to be the most natural way to couple the vacuum energy to gravity. Even if more complicated couplings are introduced, if variations of the action are determined perturbatively, this zeroth-order minimal coupling term will appear in much the same way.

This raises a few philosophical concerns regarding semiclassical treatments of gravity. There is a strong indication that classical general relativity is incompatible with quantum field theory, and semiclassical treatments of the two merged can differ depending on the route one takes. As presented above, the priority lies in defining quantities on a highly symmetric Minkowski background, and introducing gravity as a coupling between the metric and matter on a background Minkowski spacetime. In effect, this is equivalent to treating spacetime curvature due to gravity as a perturbation to a Minkowski background space. Though certainly the consensus procedure amongst theorists working on a quantum theory of gravity—and the basis for string theory—other approaches to merging general relativity and quantum field theory exist. In particular, some physicists argue that a major desideratum for quantum gravity is that, like general relativity, it should be a background-independent theory.[14] This means that a background Minkowski spacetime is not assumed, and then quantities like the vacuum energy cannot be defined as they currently are in ordinary quantum field theory (Smolin 2006). These differences may be reflected in other approaches to semiclassical gravity, where quantum fields are defined on curved spacetime backgrounds. Hollands and Wald (2010), among others, have developed an axiomatic framework for quantum field theory on curved spacetimes, and vacuum energy does not arise in this context, since a vacuum state is not in general well defined.

> One of the key insights is that—apart from stationary spacetimes or spacetimes with other very special properties—there is no unique, natural notion of a 'vacuum state' or of 'particles'. Indeed, unless the spacetime is asymptotically

[14] It's not exactly clear that general relativity is best understood as a background-independent theory, but the insight that the dynamics play an important role in dictating the spacetime structure is well established.

> stationary at early or late times, there will not, in general, even be an asymptotic notion of particle states. (Hollands and Wald 2010, 86)

Though decidedly against orthodoxy, this remains a viable avenue of research, and completely does away with the CCP by privileging curved spacetimes over a Minkowski background.

Even in the standard treatments involving minimal coupling, the use of the expectation value of the vacuum energy may be suspect. Wang et al. (2017) emphasize that a more fully quantum treatment of the vacuum energy would highlight that it is far from constant.

> Instead, at each spatial point, the spacetime sourced by the vacuum oscillates alternatively between expansion and contraction, and the phases of the oscillations at neighboring points are different. In this manner of vacuum gravitation, although the gravitational effect produced by the vacuum energy is still huge at sufficiently small scales (Planck scale), its effect at macroscopic scales is largely canceled. (1f)

If Wang et al. are correct, then the 'predicted' cosmological constant is an artifact of an improperly classical treatment of the vacuum energy in quantum field theory. An appropriately quantum treatment then leads to exponential suppression of the vacuum energy at macroscopic scales.

Though minimal coupling of $\langle \rho_{vac} \rangle$ to gravity is certainly an open option, it is by no means the only available means for coupling quantum matter to gravity. Given that it leads to such a terrible 'prediction' of the cosmological constant, alternative avenues like those just sketched should be more widely pursued. If any alternative to the orthodox coupling turns out to be correct, then the CCP is likely to disappear again.

12.2.4 Does Anything Else Act Like Λ?

If we reject the concerns raised in the previous three steps, we are left with a cosmological constant 'prediction' from quantum field theory that is at least fifty orders of magnitude larger than the observed Λ used in the ΛCDM model of cosmology. If the quantum vacuum energy is the lone source of the cosmological constant, then this is a huge problem—even a 'crisis'. However, there may be other exotic physics that leads to additional counterterms in the Einstein field equations, such that

$$\Lambda_{observed} = 8\pi G \langle \rho_{vac} \rangle + C_1 + C_2 + \cdots, \tag{7}$$

where each C_i is a contribution from some other physical mechanism. In effect, this would be a form of solution to the CCP, since the new counterterms would presumably be high-energy effects from a theory of quantum gravity. However, it seems reasonable to first treat the sources we know of; I do not see any problems in this final step, and include it only for completeness.

As stated above, if any of the arguments presented in sections 12.2.1 to 12.2.3 are accepted, then the CCP is no longer a problem. Even if the arguments are rejected, I hope to have shown that there is far more subtlety to the CCP than standard treatments suggest. Setting aside these concerns, however, in the next section I look at some broad classes of potential solutions. Even if the problem is taken seriously, these potential solutions have problems of their own.

12.3 Plausibility of Attempted Solutions

Because of the apparent attraction of the CCP as a 'veritable crisis', there are nearly as many proposed solutions as there are physicists who have heard of the problem. In this section, I will outline a few different strategies for a solution, and some problems that prevent any from standing out as the best way forward. Given the criticisms of the formulation of the problem outlined above, the lack of a clear solution strategy is further reason to reject the CCP as worth pursuing.[15]

12.3.1 't Hooft Naturalness

't Hooft naturalness, also known as technical naturalness, is a mathematical means of shielding otherwise unconstrained parameters from being too large. If a theory contains a parameter that would go to zero with an added symmetry, and that added symmetry is not too badly broken by the original theory, then the scale of symmetry breaking 'protects' that parameter from getting too large. This notion of naturalness, as the name implies, is due to 't Hooft (1980). The classic example is the pion mass due to near chiral symmetry. If chiral symmetry—the symmetry between left-handed and right-handed particles—were an exact symmetry in nature, the pseudoscalar meson masses would be forced to zero. Spontaneous breakdown of global chiral symmetry would then lead to the presence of massless Goldstone bosons. Nambu and Jona-Lasinio (1961) originally posited that the pion would be the Goldstone boson related to chiral symmetry breaking in the limit of

[15] For more in-depth reviews of different approaches to solving the CCP, review articles such as Nobbenhuis (2006) are a good place to start. Here I am concerned simply with common conceptual features of broad classes of solutions, and do not expand on any technical details.

equal mass fundamental states—what we would now call quarks. Since the quarks are *not* massless or of equal mass, neither is the pion, but the pion's relatively small mass is protected by the approximate flavor symmetry between the up and down quarks. As the bare masses of the up and down quarks approach equality, the pion mass approaches zero. Naturalness thus explains why the pion mass is so much smaller than other quark bound-states. The hope regarding the CCP is that one can find an approximate symmetry under which the vacuum energy would be zero, to help explain its small observed value in the form of a cosmological constant.

The most well-known naturalness solution is supersymmetry. Vacuum energy contributions from bosons and fermions are of opposite sign, so if each known particle in the Standard Model had a superpartner with the same quantum numbers, all contributions would cancel out, resulting in total vacuum energy of zero. Since supersymmetry must be a broken symmetry, the vacuum energy would take some nonzero value, but would be protected from divergence when supersymmetry is restored. Standard theoretical predictions require the energy scale at which supersymmetry is restored to be proportional to the square root of the residual vacuum energy, $M_{SUSY} \propto \Lambda^{1/2}$, and the lack of observed superpartners up to the currently accessible TeV scale means that supersymmetry—if it exists—is too badly broken to protect the cosmological constant.

Other symmetries have been proposed to protect the vacuum energy, including conformal symmetry. Conformal symmetry forbids any terms in a Lagrangian that set a length scale, so the vacuum energy would be excluded in a conformally invariant theory. Unfortunately for the plausibility of conformal invariance, mass terms are also disallowed. However, conformal invariance can be used as a motivation for modifying the Einstein-Hilbert action, in effect rewriting the action in terms of the trace-free component of the Riemann tensor. A residual cosmological constant term then comes out as a boundary term for the action. This leaves open the question of how the quantum vacuum should factor into gravity.

The problem with most of the approaches utilizing 't Hooft naturalness is that the symmetries would need to be far more exact empirically to dramatically reduce the magnitude of the vacuum energy as standardly predicted. Outside of supersymmetry, naturalness does not address the 'prediction' of $\langle \rho_{vac} \rangle$ from quantum field theory. If the CCP is taken seriously in its standard form, something other than naturalness seems to be needed to solve it.

12.3.2 Violate Equivalence Principle

If one believes in the vacuum energy from quantum field theory, a plausible way to dissolve the CCP is to propose a mechanism to decouple the vacuum energy from the Einstein field equations. In effect, this is a violation of the Einstein equivalence principle, since then only nonvacuum energy would gravitate. I include in this class of solutions solutions that generalize the equivalence principle as well. In one class of proposals, extra dimensions are introduced, and the large vacuum energy leads to high curvature and compactification of all but the familiar four spacetime dimensions of classical general relativity (cf. Nobbenhuis 2006, section 5.1). In this class of solutions, the equivalence principle is technically not violated, but only appears to be violated because of our ignorance of the extra dimensions. This is broadly compatible with string theories, but the bounds on the size of compactified dimensions severely limit the range of permissible string theories. Most string theories posit much larger compactified dimensions, so the addition of new submillimeter-scale dimensions is a high price to pay for a solution to the CCP, especially if we reject the arguments from the previous section that it is an ill-motivated problem. A further mechanism must also be included to explain why all but four dimensions would compactify because of $\langle \rho_{vac} \rangle$, as well as why the equivalence principle—at least approximately—holds for nonvacuum energy in four dimensions. These further problems are not solved by a simple appeal to string theory, although it is possible that work in string theory could provide answers in the future.

Further strategies involve infrared modifications to the gravitational interaction, leading to effectively massive gravitons (e.g., Arkani-Hamed et al. 2004), so-called fat gravitons (e.g., Sundrum 2004), or symmetry breaking where gravitons appear as the Goldstone bosons associated with the broken symmetry (e.g., Weinberg and Witten 1980). A benefit of infrared modifications is that small corrections to general relativity may be observable at galactic or even solar system scales, making this class of attempted solutions empirically testable. However, many of these solutions end up also violating the assumption that gravitational physics has a clean separation of energy scales. Microscopic phenomena (such as modifications to graviton couplings) end up directly altering cosmic expansion rates. As mentioned in section 12.2.2, the cutoff dependence of ρ_{vac} depends conceptually on an effective field theory framework, in which there is a clear separation of energy scales. This brings up a worry that these sorts of modifications undermine the very problem they wish to solve. To this point in time, modifications to general relativity have not been observed, and these models are therefore constrained by observation in the degree to which they can modify gravity at observable scales.

12.3.3 Statistical Solutions

Considerations from quantum cosmology and inflation have led many physicists to take seriously the prospect that our universe is just one in an ensemble of universes, making up the multiverse. It is further conjectured that the fundamental constants of nature—such as the electron charge, strong coupling constant, and the cosmological constant—are variable between universes. The ensemble is assumed to be so vast that all physically possible values of the constants are realized in some universe. We can thus put bounds on possible values of Λ with anthropic considerations (Weinberg 1989). Despite the numerous concerns involved in defining a measure over an ensemble of unobservable universes, this sort of 'solution' to the CCP does not explain why the apparently large vacuum energy from quantum field theory does not gravitate in the form of a cosmological constant.

If we look at the wavefunction of the universe, it should in general be a superposition of all physically possible values of the cosmological constant. Hawking (1984) argued that the wavefunction of the universe would be heavily peaked around $\Lambda = 0$. If correct, this might help solve the old CCP, but the new CCP becomes more puzzling in light of Hawking's work. Both the anthropic and quantum statistical solutions to the CCP suffer from the same sorts of defects. First, neither address the vacuum energy or its lack of gravitation. They aim only to make the observed Λ less surprising, and disconnect the quantum vacuum from the cosmological constant. Second, there is no good way to delimit which values of Λ are physically possible, and which aren't. So we have an ill-defined space of possibilities from which to sample. Even if that problem can be surmounted, there are no grounds for introducing a measure over the possibility space, and thus statistical arguments cannot be justified here. Norton (2010) argues that statistical arguments like these are doomed to fail in cosmology; in any case, they do little to solve the CCP.

12.4 Conclusions

I have argued against taking the CCP seriously. There are major hurdles to be overcome in (12.2.1) believing in an absolute value for the bare vacuum energy, (12.2.2) trusting the cutoff-dependent perturbative calculation of $\langle \rho_{vac} \rangle$, and (12.2.3) inserting the expectation value of the vacuum energy into the Einstein field equations. If one rejects any of these steps, then the CCP—at least as standardly presented—does not arise. Suppose that the CCP is meant to be a guiding problem in constructing a quantum theory of gravity. Then perhaps it needn't be well supported so long as it serves its heuristic purpose. If we look at the sheer number of attempts to solve the CCP, then it has undoubtedly been successful as a heuristic motivator. However, no clear solution strategy has led to major progress in developing quantum gravity. Many of the solution strategies outlined in section 12.3 are either ad hoc—their purpose is to solve the CCP, so they have been constructed to do little more than that—or inspired by solution strategies proposed independent of the CCP. Many do not even solve the CCP as traditionally stated. If successful, these 'solutions' would have nothing to say about the quantum vacuum.

The issues behind accepting the CCP highlight the reliance on heuristics in arguments in high-energy physics. Because of the lack of empirical access to high-energy phenomena, problems are formulated based on extrapolation of methods and heuristics that have been successful in low-energy particle physics. Often this is not a problem—symmetry arguments, gauge theory, and locality are generally taken to apply at all energy scales—but the heuristics seem to break down in the case of the CCP.

It is possible that a better argument supporting the CCP can be formulated. If so, it would be interesting to see the details. Unless it can avoid all of the concerns I have raised in section 12.2, I am skeptical that it should be accepted. However, even if the problem can be strengthened to avoid some of my concerns, the improved generality may help to better highlight the assumptions involved in the CCP, as well as the best way to 'solve' it in a new theory of quantum gravity.

References

Afshordi, N. (2015). Why is high energy physics Lorentz invariant? arXiv preprint arXiv: 1511.07879.

Arkani-Hamed, N., H.-C. Cheng, M. A. Luty, and S. Mukohyama (2004). Ghost condensation and a consistent infrared modification of gravity. *Journal of High Energy Physics 2004*(05), 074.

Benétreau-Dupin, Y. (2015). The Bayesian who knew too much. *Synthese 192*(5), 1527–1542.

Bianchi, E. and C. Rovelli (2010). Why all these prejudices against a constant? arXiv preprint arXiv: 1002.3966.

Burgess, C. P. (2015). The cosmological constant problem: Why it's hard to get dark energy from micro-physics. In C. Deffayet, P. Peter, B. Wandelt, M. Zaldarriaga, and L. F. Cugliandolo (Eds.), *Post-Planck Cosmology: Lecture Notes of the Les Houches Summer School: Volume 100, July 2013*, pp. 149–197. New York: Oxford University Press.

Casimir, H. B. (1948). On the attraction between two perfectly conducting plates. *Proceedings of the KNAW 51*, 793–795.

Hawking, S. W. (1984). The cosmological constant is probably zero. *Physics Letters B 134*(6), 403–404.

Hollands, S. and R. M. Wald (2010). Axiomatic quantum field theory in curved spacetime. *Communications in Mathematical Physics 293*(1), 85–125.

Jaffe, R. (2005). Casimir effect and the quantum vacuum. *Physical Review D 72*(2), 021301.

Kotzen, M. (2012). Selection biases in likelihood arguments. *The British Journal for the Philosophy of Science 63*(4), 825–839.

Lamb, W. E. and R. C. Retherford (1947). Fine structure of the hydrogen atom by a microwave method. *Physical Review 72*(3), 241.

McCoy, C. D. (2018). The implementation, interpretation, and justification of likelihoods in cosmology. *Studies in History and Philosophy of Modern Physics 62*, 19–35.

Nambu, Y. and G. Jona-Lasinio (1961). Dynamical model of elementary particles based on an analogy with superconductivity. *Physical Review 122*(1), 246–254.

Nobbenhuis, S. (2006). Categorizing different approaches to the cosmological constant problem. *Foundations of Physics 36*(5), 613–680.

Norton, J. D. (2010). Cosmic confusions: Not supporting versus supporting not. *Philosophy of Science 77*(4), 501–523.

Padilla, A. (2015). Lectures on the cosmological constant problem. arXiv preprint arXiv: 1502.05296.

Rugh, S. E. and H. Zinkernagel (2002). The quantum vacuum and the cosmological constant problem. *Studies in History and Philosophy of Modern Physics 33*(4), 663–705.

Saunders, S. (2002). Is the zero-point energy real? In M. Kuhlmann, H. Lyre, and A. Wayne (Eds.), *Ontological Aspects of Quantum Field Theory*, pp. 313–343. Singapore: World Scientific.

Smolin, L. (2006). The case for background independence. In D. Rickles, S. French, and J. Saatsi (Eds.), *The Structural Foundations of Quantum Gravity*, pp. 196–239. Oxford: Oxford University Press.

Sparnaay, M. (1957). Attractive forces between flat plates. *Nature 180*(4581), 334–335.

Sundrum, R. (2004). Fat gravitons, the cosmological constant and submillimeter tests. *Physical Review D 69*(4), 044014.

't Hooft, G. (1980). Naturalness, chiral symmetry, and spontaneous chiral symmetry breaking. In G. 't Hooft, C. Itzykson, A. Jaffe, H. Lehmann, P. K. Mitter, I. M. Singer, and R. Stora (Eds.), *Recent Developments in Gauge Theories: NATO Advanced Study Institutes Series (Series B. Physics)*, Volume 59, pp. 135–157. Boston, MA: Springer.

't Hooft, G. and M. Veltman (1972). Regularization and renormalization of gauge fields. *Nuclear Physics B 44*, 189–213.

Wang, Q., Z. Zhu, and W. G. Unruh (2017). How the huge energy of quantum vacuum gravitates to drive the slow accelerating expansion of the universe. *Physical Review D 95*(10), 103504.

Weinberg, S. (1989). The cosmological constant problem. *Reviews of Modern Physics 61*(1), 1–23.

Weinberg, S. (1999). What is quantum field theory, and what did we think it is? In T. Y. Cao (Ed.), *Conceptual Foundations of Quantum Field Theory*, pp. 241–251. Cambridge: Cambridge University Press.

Weinberg, S. and E. Witten (1980). Limits on massless particles. *Physics Letters B 96*(1–2), 59–62.

White, R. (2000). Fine-tuning and multiple universes. *Nous 34*(2), 260–276.

Wilson, K. G. and J. Kogut (1974). The renormalization group and the ε expansion. *Physics Reports 12*(2), 75–200.

Index

Note: Page numbers followed by '*f*' or '*t*' refer to figures and tables.